Computational welding mechanics

Related titles:

Fatigue assessment of welded joints by local approaches – Second edition
(ISBN 978-1-85573-948-2)
Local approaches to fatigue assessment are used to predict the structural durability of welded joints, to optimise their design and to evaluate unforeseen joint failures. This completely reworked second edition of a standard work provides a systematic survey of the principles and practical applications of the various methods. It covers the hot-spot structural stress approach to fatigue in general, the notch stress and notch strain approach to crack initiation and the fracture mechanics approach to crack propagation. Seam-welded and spot-welded joints in structural steels and aluminium alloys are also considered.

Fatigue analysis of welded components: Designer's guide to the hot-spot stress approach
(ISBN 978-1-84569-124-0)
This report provides practical guidance on the application of the structural hot-spot stress approach to promote wider use of the approach in the development of better fatigue design and analysis procedures for welded structures. The guide is also intended to assist code-writers and to encourage further research into design standards.

Cumulative damage of welded joints
(ISBN 978-1-85573-938-3)
Fatigue is a mechanism of failure that involves the formation of cracks under the action of different stresses. To avoid fatigue it is essential to design a structure with inherent fatigue strength. However, fatigue strength is not a constant material property and any calculations are necessarily built on a number of assumptions. *Cumulative damage of welded joints* explores the wealth of research in this important area and its implications for the design and manufacture of welded components.

Details of these and other Woodhead Publishing materials books, as well as materials books from Maney Publishing, can be obtained by:

- visiting our web site at www.woodheadpublishing.com
- contacting Customer Services (e-mail: sales@woodhead-publishing.com; fax: +44 (0) 1223 893694; tel.: +44 (0) 1223 891358 ext. 130; address: Woodhead Publishing Limited, Abington Hall, Abington, Cambridge CB21 6AH, England)

Maney currently publishes 16 peer-reviewed materials science and engineering journals. For further information visit www.maney.co.uk/journals.

Computational welding mechanics

Thermomechanical and microstructural simulations

Lars-Erik Lindgren

Woodhead Publishing and Maney Publishing
on behalf of
The Institute of Materials, Minerals & Mining

CRC Press
Boca Raton Boston New York Washington, DC

WOODHEAD PUBLISHING LIMITED
Cambridge England

Woodhead Publishing Limited and Maney Publishing Limited on behalf of
The Institute of Materials, Minerals & Mining

Published by Woodhead Publishing Limited, Abington Hall, Abington,
Cambridge CB21 6AH, England
www.woodheadpublishing.com

Published in North America by CRC Press LLC, 6000 Broken Sound Parkway, NW,
Suite 300, Boca Raton, FL 33487, USA

First published 2007, Woodhead Publishing Limited and CRC Press LLC

British Library Cataloguing in Publication Data
A catalogue record for this book is available from the British Library.

Library of Congress Cataloging in Publication Data
A catalog record for this book is available from the Library of Congress.

Woodhead Publishing Limited ISBN 978-1-84569-221-6 (book)
Woodhead Publishing Limited ISBN 978-1-84569-355-8 (e-book)
CRC Press ISBN 978-1-4200-6337-0
CRC Press order number WP6337

The publishers' policy is to use permanent paper from mills that operate a sustainable forestry policy, and which has been manufactured from pulp which is processed using acid-free and elementary chlorine-free practices. Furthermore, the publishers ensure that the text paper and cover board used have met acceptable environmental accreditation standards.

Project managed by Macfarlane Production Services, Dunstable, Bedfordshire, England
(macfarl@aol.com)
Typeset by Godiva Publishing Services Limited, Coventry, West Midlands, England

Contents

Foreword

The overall aim of computational welding mechanics (CWM) is to establish methods and models that are usable for control and design of welding processes to obtain appropriate mechanical performance of the welded component or structure. CWM is therefore concerned with subjects ranging from modelling of heat generation and weld pool phenomena to thermal stresses and large, plastic deformations of welded structures or components. It also includes processes such as stress relief heat treatment, which are applied in conjunction with welding. The book will be of considerable use for anyone concerned with manufacturing simulations as thermomechanics and changing material microstructure are not unique for welding, although they are more pronounced. Meeting the CWM challenge will empower its students for simulation of manufacturing processes in general.

The book will be of interest for managers and project leaders who want to know what resources are needed to apply the tools of CWM and what competence and improvement of processes/products their application can lead to. Furthermore, engineers will be given detailed information in order to understand phenomena and design issues as well as guidelines for modelling and simulation. Researchers, doctoral students and lecturers will find the state of the art, theoretical foundations and useful numerical methods described. The latter are demonstrated by the accompanying software implemented in Matlab™. This consists of a number of utility programs and one-dimensional finite element codes for specific demonstrations as well as a complete program for coupled thermo-metallurgical-mechanical simulation of welding problems. The latter program is applicable for three-dimensional problems using 3D solid elements. It also includes addition of filler material and mixture rules for material properties.

Preface

Writing this book has been like a journey through my own experience in computational welding mechanics (CWM) as well as through others' experience via the research literature and private communications. The outcome has not only been a book where I have been content to share my experience and expertise in this field. I have also learned new things when delving into some of the details and even more when formulating strategies for calibration, validation and modelling. The keyword during the writing of the book has been 'enabling'. Therefore the book has somewhat of a textbook style and contains demonstrations of the theory through Matlab^{TM} programs. The book contains considerably more than may be apparent from its size. The CWM_Lab software comprises a number of programs for demonstration of different aspects in CWM. It has been a personal satisfaction to learn more about the most modern hyperelasto-plastic formulations when implementing this into the finite element program, CWM_Lab, for thermo-metallurgical-mechanical problems (see pages xiii–xiv). Therefore, the book also demonstrates theory in a larger context of finite element methods and not only CWM.

I do hope that the book, together with its software, will be of benefit to anyone that wants to master CWM or simulation of manufacturing processes in general. I acknowledge the initiative from Professor Dieter Radaj that led to the writing of this book. His experience in authorship as well as in CWM has been of great help to me.

I would like to express my thanks to Professor Lennart Karlsson, who brought me into this field of research and to Professor Mikael Jonsson who pursued his PhD studies in CWM with me. I want to acknowledge the financial support for my research in CWM over the past two decades from VINNOVA, the Swedish Governmental Agency for Innovation Systems, and NFFP, National Aviation Research Programme together with Volvo Aero Corporation. I would like to mention Professor Bengt-Olof Elfström at Volvo Aero Corporation. His vision about 20 years ago of being able to simulate all manufacturing processes at their company has been important for these achievements (see Chapter 13). I also gratefully acknowledge the contributions from previous and current PhD students in CWM: Henrik Alberg, Bijish Babu, Daniel Berglund, Lars Börjesson, Erika Hedblom, Andreas Lundbäck and Henrik Runnemalm. Furthermore, I want to extend my thankfulness to colleagues all over the world and especially Professor John Goldak and Professor Moyra McDill at Carleton University, Ottawa.

Finally I would like to dedicate this book to my personal friend and colleague Dr Alan Oddy, who passed away too early in March 2001. Several ideas in this book come from what I learned from him and through his publications. I do miss our stimulating talks about problems and solutions in CWM but even more his sense of humour, as well as his friendship with me and my family.

Lars-Erik Lindgren
Luleå, Sweden

CWM_Lab software

Some of the numerical methods described in this book are illustrated by examples using CWM_Lab software. This software, implemented in MATLAB^{TM}, is a finite element program for solving coupled thermo-metallurgical-mechanical welding problems. The program is applicable for three-dimensional problems using 3D solid elements. It also includes addition of filler material and mixture rules for material properties.

The CWM-Lab program is available as downloads from the Internet by e-mailing the author at: lel@ltu.se. The software is free for non-commercial use, e.g. for educational and research purposes. Special conditions apply for commercial use of the software. A password allowing access to the software and instructions on its use will be sent by e-mail on receipt of the following:

- the name and affiliation of the person who wishes to use the software;
- confirmation whether the software will be used for commercial or non-commercial purposes;
- a copy of an invoice or receipt confirming purchase of this book. The copy should be sent as a PDF scan attached to the e-mail requesting access to CWM_Lab.

A flowchart of the main functions in CWM_Lab is shown in Fig. 1. Some of these functions in CWM_Lab are referred to in the text. These are:

- **LEL_Thermod**. A routine for iterative solutions in thermal analysis.
- **LEL_ther_El_Loop**. A function in LEL_Thermod that loops over all thermal elements.
- **LEL_Mechmod.** A routine for assembling right-hand side and tangent stiffness.
- **LEL_Mech_Loop.** A routine for iterative solutions in mechanical analysis.
- **LEL_RadialReturn.** A routine for assembling right-hand side and tangent siffness.
- **LEL-TransPlast.** A routine to compute deviatoric transformation plasticity.
- **LEL-Element.MPC.** A function in LEL_Mechmod for modifying element force and matrix for multipoint constraint (MPC).

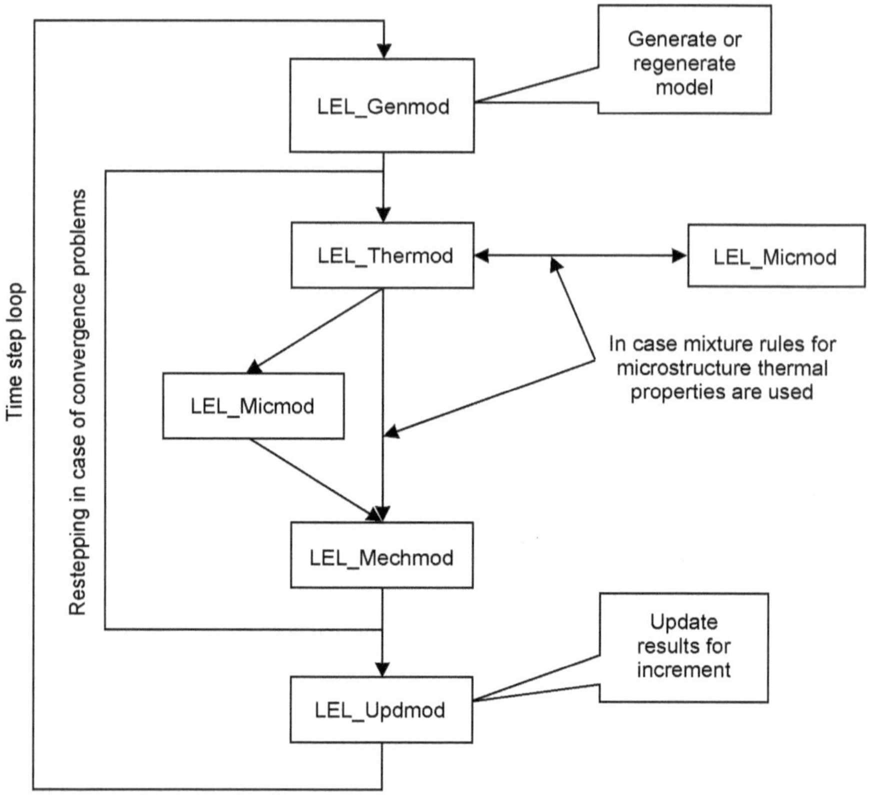

1 Flowchart of CWM_Lab.

- **LEL-DoubleEllipsoidPowerWeld.** A function in LEL_Thermod which integrates the power from a weld source with the heat source as a double ellipsoid Gaussian distribution.

Boxed examples in the text illustrate methods to solve particular problems. A number of these use other MATLAB™ functions. These codes are also available as downloads. The boxed examples using these codes are:

- Box 3.1: Box_function 1 LEL_Box_CoupledExample.
- Box 3.3: Box_function 2 LEL_CoupledLinearThermoElasticityExample.
- Boxes 3.4 and 3.5: Box_function 3 LEL_Box_Heat_Cond_ConvExample.
- Box 5.1: Box_function 4 LEL_Box_Rod_Melting.
- Box 6.6: Box_function 5 LEL_Box_Pure_Shear.
- Box 9.1: Box_function 6 LEL_Box_TTT_diagram.
- Box 9.2: Box_function 7 LEL_Box_Heatsource.
- Box 11.1: Box_function 8 LEL_Box_Buckling_Rigid_Box.

1 Introduction

1.1 Computational welding mechanics

The overall aim of computational welding mechanics (CWM) is to establish methods and models that are usable for control and design of welding processes to obtain appropriate mechanical performance of the welded component or structure. It is therefore concerned with subjects ranging from modelling of heat generation and weld pool phenomena, heat flow to thermal stresses and deformations. Material science and constitutive modelling are essential ingredients in the modelling of welding processes due to the severe thermal cycle(s) during welding. CWM models can be combined with models for microstructure evolution and other features that enable the prediction of microstructure, cracking and other phenomena that are determined by the temperature and deformation history of the material.

The centrepiece in welding simulations is the heat generation process. Its description belongs to the domain of thermomechanics in the case of explosive welding, friction welding and friction stir welding. An electrical field is also needed in resistance welding. However, the process becomes much more complex for fusion welding processes. Weld process modelling (WPM) focuses on modelling the physics of heat generation. CWM models, on the other hand, start with a given heat input that replaces the details of the heat generation process and focus on larger scales. The modelling of fluid flow and pertaining convective heat transfer may be integrated with a CWM model. However, the classical approach in CWM is to ignore fluid flow and use a heat input model where heat distribution is prescribed. Thus the heat input model in CWM must be calibrated with respect to experiments or obtained from WPM models. Therefore, the classical CWM models have some limitations in their predictive power when used to solve different engineering problems. For example, they cannot prescribe what penetration a given welding procedure will give. The appropriate procedure to determine the heat input model is therefore important in CWM.

The use of computational models does not replace experimental methods but does mean their role is redesigned. Fewer experiments (tests), are needed to evaluate different design concepts when the power of computer models is applied. Furthermore, the more established simulations become in a given field, the less validation testing is needed. However, more demands are placed on determination of material properties and boundary conditions needed for the computational model.

1.2 Contents of the book

The scope of this book is limited to modelling the thermomechanics of fusion processes of metals. The inclusion of pre- and post-weld heat treatment procedures in the simulation chain is quite natural in this context. The models and methods described in the book can also easily be applied to these procedures. Some examples of simulations where post-weld heat treatment is simulated in conjunction with welding simulations are discussed and the extension of the methods to thermal cutting is also straightforward [1]. The differences between various fusion welding processes in CWM modelling are small. They are accounted for in the heat source models by varying the distribution of energy and the way the addition of the filler material is done.

This introductory part is followed by a description of the multi-physics of welding (Chapter 2), and how to simplify this (Chapter 3), and an illustration of the thermomechanical phenomena to be aware of when modelling welding processes (Chapter 4). Thereafter, Chapters 5 and 6 describe the finite element formulations typically used in CWM. There is more about numerical techniques and modelling techniques for efficient simulations in Chapter 7. In Chapter 8, a methodology for calibration and validation of the models is outlined, followed by a discussion about modelling options in Chapter 9 and a proposal for modelling strategy in Chapter 10. Chapter 11 is concerned with the problem of robustness and stability of a welding process. Chapter 12 provides a short update of earlier reviews of work discussing the improvement of the practice of CWM as well as applications of CWM. Thereafter, the story of successful industrial application of CWM by Volvo Aero is given. The book is closed with the conclusions in Chapter 14.

1.3 The competent company

Competition in a global market drives each company to be efficient and produce products of high quality. 'Time to the market' and 'Right the first time' are common expressions for this. Combined with development in computational tools and computer hardware, this has led to an increase in the use of tools for virtual design. The finite element method has become an indispensable tool in product development [2].

The use of manufacturing simulations is less common because of the difficulties in applying them. However, the problem is not due to a lack of

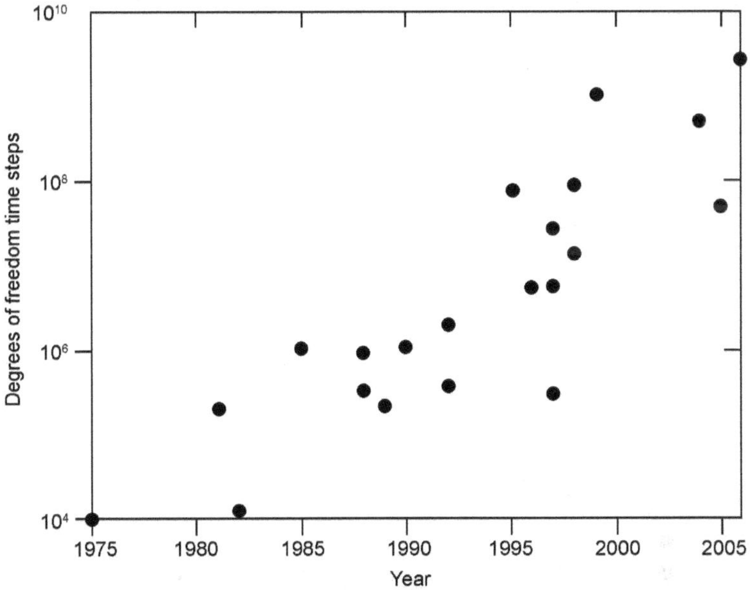

1.1 Size of CWM models. The size is measured as the number of degrees of freedom in the mechanical part of the finite element model multiplied by the number of time steps.

computing power, as is illustrated in Fig. 1.1. The bottleneck for increasing use of advanced simulations is the need for competence! Simulations require competence. This is not only a competence in the use of advanced numerical methods but even more the required quantitative knowledge about processes, materials, etc., needed to set up a model. This requirement is in itself a positive requirement, enforcing the company to understand its manufacturing processes in order to be able to improve them. However, the simulations do not only require competence, they also give competence.

A successful company continuously adapts and develops its internal processes for design, manufacturing, marketing and customer support. The management of human resources and information technology (Fig. 1.2) is crucial in this. Information technology can be used for acquiring, storing and communicating knowledge. Product data management (PDM) systems, product information management (PIM) systems and product lifecycle management (PLM) tools are information technologies used to support product development and also to facilitate teamwork. There is also a need to manage the design process itself. This is done by establishing guidelines for the development of models and their use, together with design criteria. It is preferable to formulate this expertise in written format and also possible to implement it into tools for knowledge-enabled engineering (KEE). These tools can also be combined with systems for model data management and team collaboration.

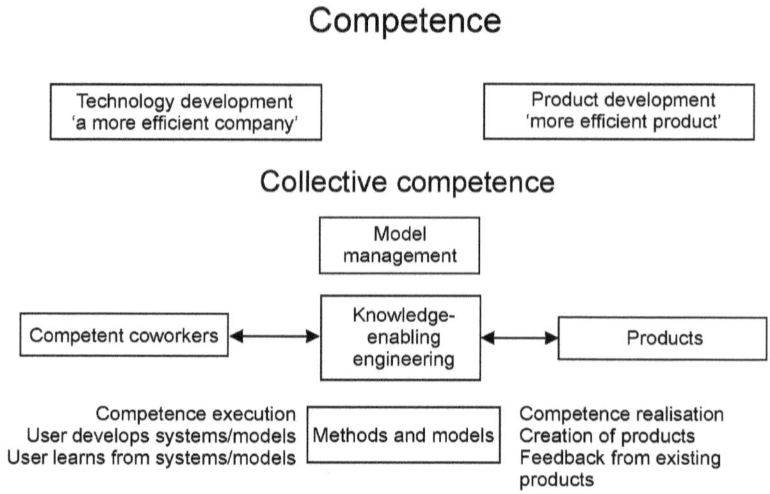

1.2 The competent company.

CWM belongs to the category of 'Methods and models' in Fig. 1.2. Typically, CWM is used in aerospace applications, nuclear power plants and the automotive industry where safety and quality are important. The Volvo Aero story in Chapter 13 illustrates this first case and Volvo Aero's route to world-class competence in CWM stems from a strategic choice made around 1990. Other companies need results from CWM less often and then outsource the simulations. It is then important to have enough competence in CWM to communicate with the consultant. One example of the latter is a multipass weld case, described by Lindgren *et al.* [3], where results from CWM were used by a small company to include welding in an application, which reduced their costs considerably.

1.4 Driving forces for increased use of welding simulations

The finite element method is the overall dominant tool used in CWM. The basic equations and finite element formulations, including nonlinear issues, are discussed in Chapters 3, 5 and 6. Although simulations do not replace experiments, as stated earlier, they do offer a number of advantages. It is possible to:

- perform virtual experiments where all parameters can be fully controlled;
- visualise the process;
- design and optimise the welding procedure, fixtures, etc.;
- use the computed fields for subsequent analyses of risk for cracking, etc., or they can be part of the simulation of a manufacturing chain (see Chapter 13).

Welding procedure specifications (WPS) ensure that the quality of a given weld procedure is qualified. This focuses on the integrity of the weld. Test pieces are welded and then checked for defects and cracks. Different set-ups are also used to create more severe thermal stresses in order to check for the risk of cracking. For example, hot cracking can be evaluated by the Varestraint and Transvarestraint test.[1]

A large number of destructive and non-destructive testing methods can be applied to ascertain the quality of welded components. The use of simulations in addition to these tests enables the evaluation both of residual stresses that affect the life of the welded component as well as of deformations that may hamper the functionality of the component. Simulations become cheaper than testing because of the reduction in cost in computer hardware. This has enabled more complex and therefore more real engineering applications to be simulated (Fig. 1.1).

1. ISO/TR 17641-3:2005, to be found at http://www.iso.org/iso/en/

2
The multi-physics of welding

Modelling of welding processes is an inherent multi-physics problem. The paper by Zacharia *et al.* [4] divided the modelling of this problem into four domains:

- heat and fluid flow;
- heat source-metal interactions;
- weld solidification microstructures;
- phase transformations.

Akhlagi and Goldak [5] classified analytic heat input models as first generation models and the types of heat input models discussed in Section 9.12 as second generation models. The higher generation models are a different kind of multi-physics models where each higher generation includes more of the physics of the problem. The book by Radaj [6] focuses on weld process models, and there is also considerable useful information in his more general book [7].

One example of different phenomena involved in welding is shown in Fig. 2.1. The model must account for high-pressure arc physics, fluid flow, heat transfer and conduction, and deformation. The model set-up for energy calculations in an arc [8] is shown in Fig. 2.1. Heat conduction, convection and radiation are included in this modelling together with the contribution to energy generation from various phenomena such as Joule heating and the Thompson effect [9].

Advanced models for predicting energy distribution coupled to the fluid flow are available [10–14]. This kind of model includes the fluid flow with driving forces, as illustrated in Fig. 2.2. The prediction of weld pool shape may even include the effect of redistribution of surfactant elements [12, 13].

Weld process models (WPM) that can predict heat generation and distribution in the weld pool include the physics of the problem and couple this with welding process parameters. One excellent example of this kind of model is by Sudnik and coworkers. The model for laser welding accounts for energy transport, equilibrium at the free surfaces for balancing vapour pressure and capillary pressure, together with a correction term. The correction term allows the metal volume to be balanced by taking into account thermal

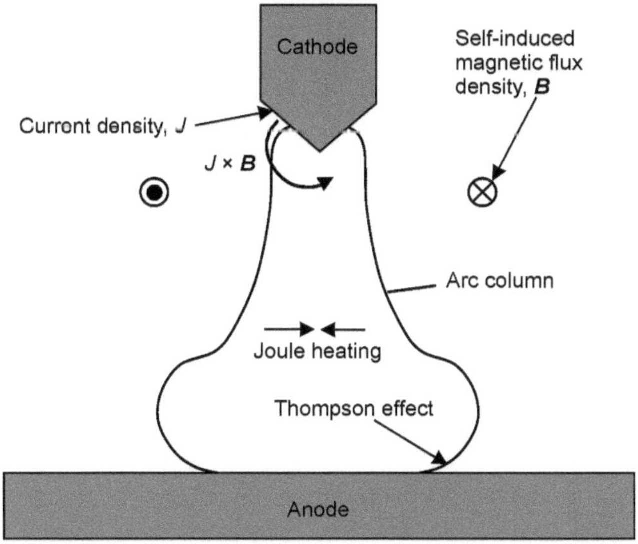

2.1 High pressure arc model for flame where weld is considered rigid, adapted from Ramirez *et al.* [8].

expansion and shrinkage during welding and the gap width between the parts to be joined [15, 16] (see Fig. 2.3). Hydrogen diffusion has been included in papers when hydrogen-induced cracking has been of interest. A discussion about hydrogen and carbon diffusion can be found in Akhlagi and Goldak [5].

The feasibility of using CWM simulations in the industrial context is due to

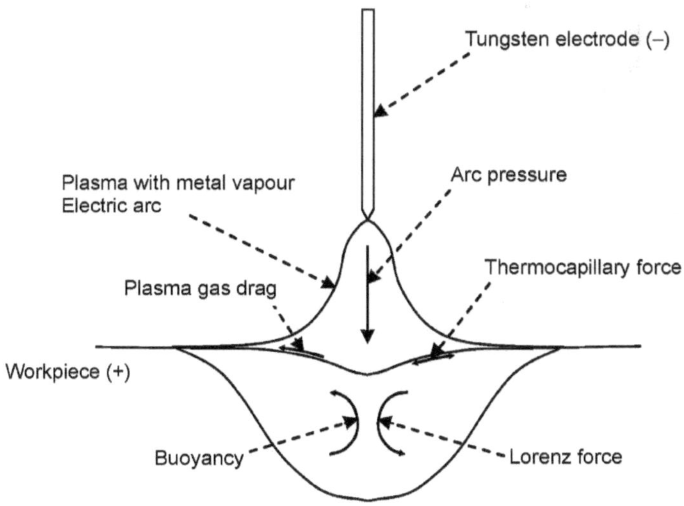

2.2 Driving forces in fluid flow in weld pool including model of plasma, adapted from Pavlyk and Dilthey [11, 14].

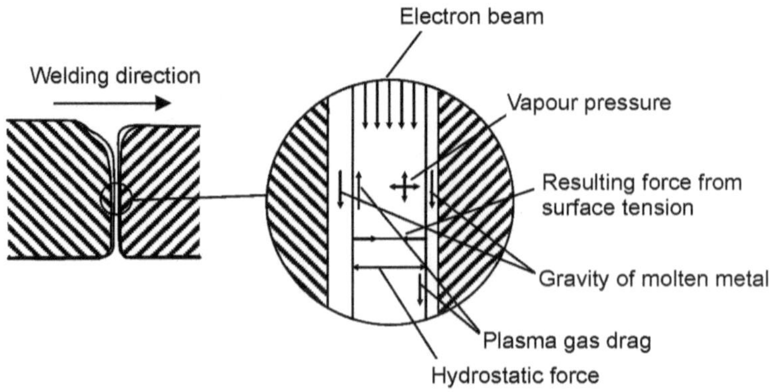

2.3 Driving forces in fluid flow in weld pool (Pavlyk and Dilthey [14]).

the possibility of reducing the multi-physics problem to weakly coupled thermomechanical models (Section 3.3). Heat generation is then determined from experiments or experience (Section 8.5.2). The definition of CWM in Section 1.1 would include weld process models of the types indicated above when they are used to predict the mechanical performance of the welded component or structure. However, there are currently no models that integrate all these phenomena together with the deformation behaviour of the welded component. The weld process models can be combined with or replace experimental results when calibrating heat sources [17]. There are only a few papers that include fluid flow together with thermomechanics [18, 19].

3
Couplings and reference frames

As illustrated in the previous chapter, welding is a multi-physics problem where the physical phenomena are described by different coupled field equations that overlap or have a common boundary. Fortunately, many welding processes can be represented by simplified models. The main stream in CWM is the use of weakly coupled models where the physics in the weld is replaced by a heat input model. A general description about procedures for solving coupled systems is given in this chapter with focus on the use of the so-called staggered approach common in CWM, where the solution of the problem is split into a thermal and a mechanical phase. There are several options available in the thermal and mechanical analyses. One common concern is the choice of coordinate system. Most models use a fixed coordinate system. However, the moving heat source with near stationary conditions can be favourably treated by a moving coordinate system in some cases. This is described at the end of this chapter with references to papers in CWM using the Eulerian, moving, coordinate system. The next two chapters about finite element formulation for solution of the thermal and mechanical fields focus on the use of a fixed coordinate system.

3.1 Coupled systems and solution procedures

A large number of coupled systems exist and they are coupled in different ways [20]. There is the coupling between different components of a system with a common interface through which they interact, e.g. a fluid and solid boundary. Another type of coupling is the coupling between different physical phenomena in the same material, e.g. the coupling between temperature and deformation in the solid material.

Monolithic or partitioned treatment of the coupled systems is possible. The solution is advanced simultaneously in time in a monolithic treatment, whereas the field models are solved and advanced separately in time in a partitioned simulation. The latter is more common, has implementation advantages and can also be computationally more efficient. An example of partitioned analysis of a coupled system, taken from the papers by Felippa *et al.* [20, 21], is the two-field problem

$$\begin{cases} 3\dot{x} + 4x - y = f(t) \\ \dot{y} + 6y - 2x = g(t) \end{cases} \qquad [3.1]$$

which, when discretised in time using a backward Euler algorithm, gives

$$\begin{bmatrix} 3+4h & -h \\ -2h & 1+6h \end{bmatrix} \begin{bmatrix} {}^{n+1}x \\ {}^{n+1}y \end{bmatrix} = \begin{bmatrix} h^{n+1}f + 3^{n}x \\ h^{n+1}g +^{n} y \end{bmatrix} \qquad [3.2]$$

where h is the length of the time step and the left superscript is a time step counter. The initial conditions give the starting values needed for the algorithm. The solution of Eq. [3.2] is a monolithic treatment of the problem.

The basic idea for a staggered solution procedure is shown in Fig. 3.1. The solution is advanced in two separate steps, denoted Ax and Ay in the figure. A predictor, denoted P, is used when advancing the first field, x in this case. This is then substituted into the solution of the second field. This substitution is trivial and denoted S in the figure. The graph for the solution procedure (right part in Fig. 3.1) illustrates the logic and is usually drawn without the dashed lines. The approach can then be drawn as in Fig. 3.2, which explains why it is called a staggered approach. The staggered approach can be combined with a full step correction as illustrated in Fig. 3.3 to compute an improved predictor value ${}^{P}y$,

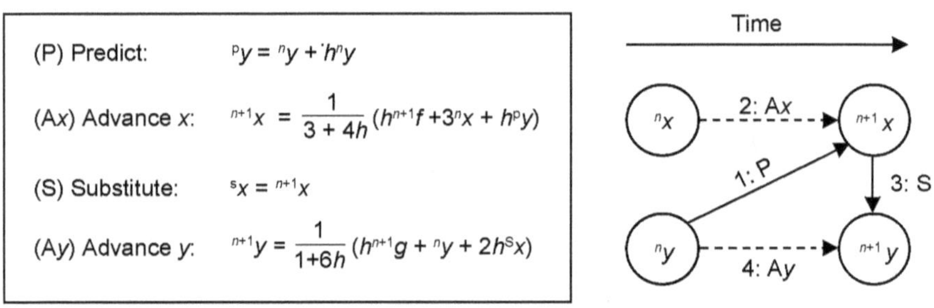

3.1 A staggered solution procedure [20]. The graph to the right illustrates the steps explained in the box to the left.

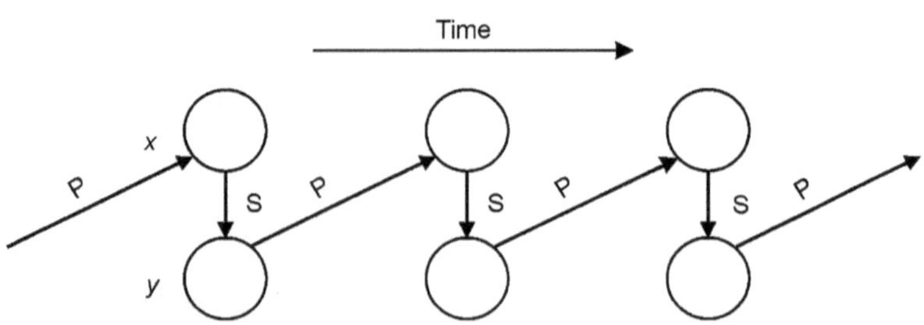

3.2 A simplified version of the graph in Fig. 3.1 typically used to illustrate the staggered solution of a coupled problem.

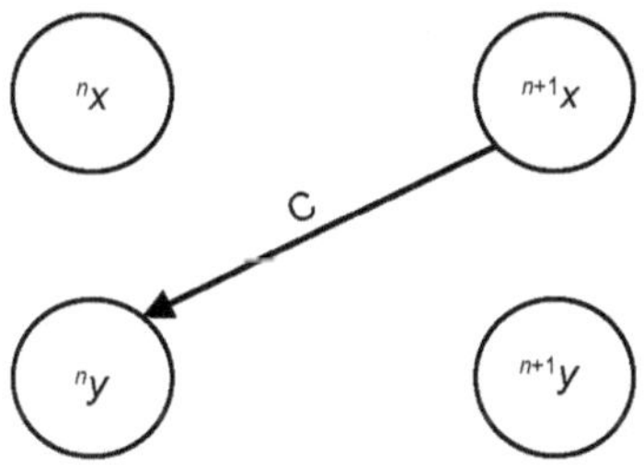

3.3 A correction step in the solution procedure [20].

and the solution according to Fig. 3.1 is repeated once again. Box 3.1 compares the fully coupled solution and the staggered approach for solving Eq. [3.2].

Other improvements of the staggered approach include the use of augmentation methods [20, 21] where the equations for advancing the solution of one field are reformulated with information from the other field. This is illustrated in Section 3.2 where an augmentation method for thermoelasticity in [20] is related to the operator split approach in [22, 23].

Box 3.1 Demonstration of solution procedures for coupled systems in Eq. [3.1]

Maple™ is used to get the analytic solution of the coupled system of equations in Eq. [3.1], using the initial conditions $x(0) = y(0) = 0$ and $f(t) = \sin(t), g \equiv 0$. The solution is

$$\begin{cases} x(t) = \dfrac{1}{845}\left[\dfrac{1}{6}r_3 r_1 e^{-1/3\bar{r}_2 t} + \dfrac{1}{6}\bar{r}_3 \bar{r}_1 e^{-1/3 r_2 t} - 113\cos(t) + 136\sin(t)\right] \\ y(t) = \dfrac{1}{845}\left[r_1 e^{-1/3\bar{r}_2 t} + \bar{r}_1 e^{-1/3 r_2 t} - 44\cos(t) + 38\sin(t)\right] \end{cases}$$

where the roots are

$$r_1 = 22 + \frac{37}{11}\sqrt{55}, \bar{r}_1 = 22 - \frac{37}{11}\sqrt{55}, r_2 = 11 + \sqrt{55}, \bar{r}_2 = 11 - \sqrt{55},$$

$$r_3 = 7 + \sqrt{55}, \bar{r}_3 = 7 - \sqrt{55}$$

The analytic solution is compared with the fully coupled solution of Eq. [3.2] and the solution using the staggered approach shown in Fig. 3.1. The difference between the two methods is quite small even when taking large time steps. The circles lie almost on the dashed lines. This can be seen in Fig. 3.4 where it can also be noted then that they both deviate considerably from the exact solution. Thus the requirement for accuracy is more

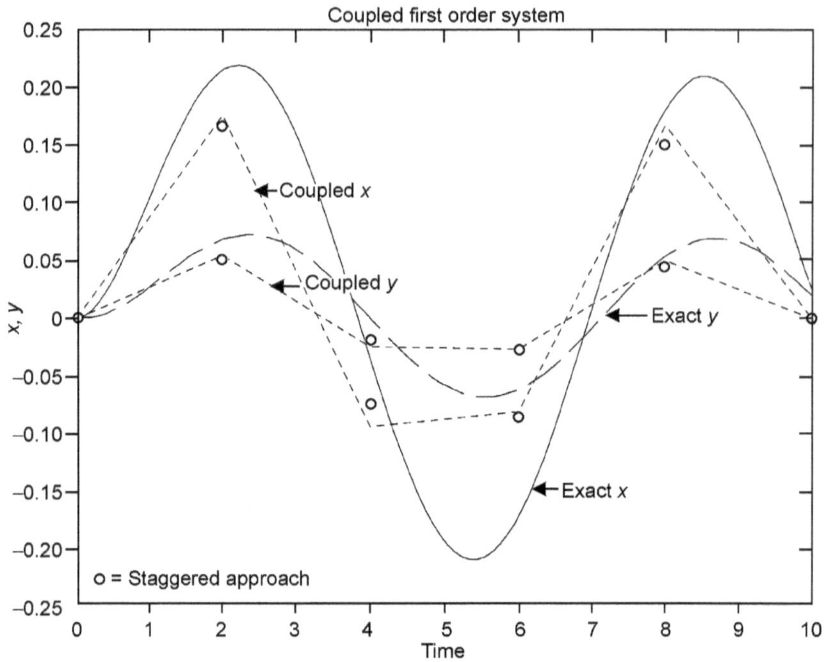

3.4 Computed *x* and *y* versus time using a time step of 2. Circles denote solution from staggered approach where as the dashed line is based on the coupled solution.

important in this particular case. An accurate solution will require smaller time steps and then the staggered approach will be more efficient than the fully coupled approach.

3.2 Linearised coupled thermoelasticity

The different solution procedures for fully coupled thermomechanical problems are illustrated by an example taken from Armero and Simo [23]. The finite element formulas below are general, although the particular case is a one-dimensional problem.

3.2.1 One-dimensional thermoelasticity

The equation of motion for a uniaxial stress case without body forces is

$$\frac{d\sigma}{dx} = \rho \ddot{u} \qquad [3.3]$$

where σ is the stress, u the displacement, $\ddot{u}$ the acceleration and ρ the density. The material behaviour for the linear thermoelastic material is taken as

$$\sigma = E(\epsilon - \alpha T) \tag{3.4}$$

where E is Young's modulus and α is the coefficient of thermal expansion. It is assumed that the body is stress free at $T = 0$. Thus the temperature can be considered as the increase in temperature from the stress free state. Furthermore, we assume small strains, ϵ,

$$\epsilon = \frac{\mathrm{d}u}{\mathrm{d}x} \tag{3.5}$$

The three equations above can be used to derive the one-dimensional Navier's equation, the equilibrium equation expressed in terms of the displacements,

$$E\frac{\mathrm{d}^2u}{\mathrm{d}x^2} - E\alpha\frac{\mathrm{d}T}{\mathrm{d}x} = \rho\ddot{u} \tag{3.6}$$

where the second term on the left hand side can now be interpreted as a volumetric load due to the thermal strains.

The energy balance in one-dimensional heat conduction can be written as

$$-\frac{\mathrm{d}q}{\mathrm{d}x} = \rho c_{\mathrm{v}}\dot{T} + E\alpha T\dot{\epsilon} \tag{3.7}$$

where q is the heat flux and c_{v} is the heat capacity at constant volume. There is some variation in the use of the latter. Some references, for example [24, 25], use c_{p}, which is the heat capacity at constant pressure. Others [26–29] use c_{v}. This variation in notation was noticed by Carslaw and Jaeger [30]. The difference between these two quantities is negligible for solids and liquids. The notation c is used in the following. The equation includes mechanically generated heat [25, 27, 28].

Fourier's law for heat flow gives

$$q = -\lambda g \tag{3.8}$$

where λ is the heat conductivity, and the temperature gradient, g, is computed as

$$g = \frac{\mathrm{d}T}{\mathrm{d}x} \tag{3.9}$$

These three equations give Duhamel's heat conduction equation

$$\rho c\dot{T} + E\alpha T_{\mathrm{a}}\dot{\epsilon} = \lambda\frac{\partial^2 T}{\partial x^2} \tag{3.10}$$

where T_{a} is the absolute temperature in K. The temperature T is the temperature difference with respect to the stress-free state according to Eq. [3.4], which can be given in different units, for example °F or °C. The equation is linearised to

$$\rho c\dot{T} + E\alpha T_{\mathrm{ref}}\dot{\epsilon} = \lambda\frac{\partial^2 T}{\partial x^2} \tag{3.11}$$

where the left hand side is equal to the entropy change. The equation is linearised by assuming $\rho T\dot{s} \approx \rho T_{\text{ref}}\dot{s}$, where s is the entropy. Thus T_{ref} must also be given in K. The second term in the equation above can be interpreted as the volumetric heat generation due to the deformation.

Boley and Weiner [29] discussed the importance of the elastic coupling term. It is related to the material properties as well as the loading. One example is the case used in Section 3.2.7 where the influence of the coupling on the wave propagation is small for short lengths of time. However, the thermal dissipation will be important if the long-term behaviour is of interest. Boley and Weiner [29] indicated that for thermoelastic cases when the inertia term in Eq. [3.6] could be neglected, the thermoelastic coupling term in Eq. [3.7] may also be neglected.

Box 3.2 Adiabatic elastic modulus

Adiabatic deformation is a deformation that is so rapid that the mechanically generated heat does not conduct significantly under the time frame of interest. Then the heat conduction equation becomes

$$\rho c\dot{T} + E\alpha T_{\text{ref}}\dot{\epsilon} = 0 \qquad [3.12]$$

The temperature increase can be computed directly from this expression and there is no need to solve a field equation. This can be useful when studying a wave propagation problem and need to account for the effect of temperature change on mechanical properties in a simple way.

Integrating the equation above with respect to time from initial stress-free state at $T = 0$ leads to

$$T = -\frac{E\alpha T_{\text{ref}}}{\rho c}\epsilon = -\frac{E\alpha T_{\text{ref}}}{\rho c}\frac{\mathrm{d}u}{\mathrm{d}x} \qquad [3.13]$$

This can then be inserted into Eq. [3.6] and, assuming homogeneous material properties, leads to

$$E\left(1 + \frac{E\alpha^2 T_{\text{ref}}}{\rho c}\right)\frac{\mathrm{d}^2u}{\mathrm{d}x^2} = \rho\ddot{u} \qquad [3.14]$$

This can be rewritten to

$${}_{ad}E\frac{\mathrm{d}^2u}{\mathrm{d}x^2} = \rho\ddot{u} \qquad [3.15]$$

where

$${}_{ad}E = E\left(1 + \frac{(E\alpha)^2 T_{\text{ref}}}{E\rho c}\right) \qquad [3.16]$$

is the adiabatic elastic modulus.

3.2.2 Finite element formulation of coupled thermoelasticity

The finite element formulation of the general deformation and heat conduction terms of the one-dimensional problem defined by Eqs [3.6] and [3.11] can be found in standard textbooks [24, 31–34]. See also Chapter 5 for nonlinear heat conduction and Chapter 6 for nonlinear deformation. Expressions for the finite element matrices and vectors for the thermomechanical coupling terms are given in Hsu [25] and Sluzalec [27]. The formulas for the coupling terms are given below, and the reader is referred to the standard textbooks for the expression for the element mass and stiffness matrices as well as for the heat conductivity and heat capacity matrices. The expressions for linear, two-node, elements can also be found in the Matlab software for Box 3.3. The finite element formulation of the coupling terms in Eqs [3.6] and [3.11] can be derived by treating them as volumetric loads.

The consistent thermal element load vector due to the mechanical generated heat in Eq. [3.6] is

$$\begin{aligned}\dot{\boldsymbol{q}}_{\mathrm{th}} &= \int_{v^{(e)}} \boldsymbol{N}_{\mathrm{th}}^{T}(-E\alpha T_{\mathrm{ref}}\dot{\epsilon})\mathrm{d}v = -\int_{v^{(e)}} \boldsymbol{N}_{\mathrm{th}}^{T} E\alpha T_{\mathrm{ref}} \boldsymbol{B}_{\mathrm{m}} \dot{\boldsymbol{u}} \mathrm{d}v \\ &= \int_{v^{(e)}} \boldsymbol{N}_{\mathrm{th}}^{T} E\alpha T_{\mathrm{ref}} \boldsymbol{B}_{\mathrm{m}} \mathrm{d}v \dot{\boldsymbol{u}} \end{aligned} \qquad [3.17]$$

where $\dot{\boldsymbol{u}}$ is the element vector with nodal velocities and $\boldsymbol{N}_{\mathrm{th}}$ is the shape function matrix interpolating the temperature field in the element from element nodal temperatures $\boldsymbol{T}^{(e)}$. The matrix $\boldsymbol{B}_{\mathrm{m}}$ gives the strain rate in the element from the nodal velocities. It contains the derivatives of the shape functions in the matrix $\boldsymbol{N}_{\mathrm{m}}$. The latter is used for interpolating the displacement field in the element from nodal values. The same shape functions for the thermal and mechanical fields in the element are used below. The coupling matrix for a two-node element becomes

$$\boldsymbol{m}_{\mathrm{u}}^{(e)} = -\int_{v^{(e)}} \boldsymbol{N}_{\mathrm{th}}^{T} E\alpha T_{\mathrm{ref}} \boldsymbol{B}_{\mathrm{m}} \mathrm{d}v = \frac{EA\alpha T_{\mathrm{ref}}}{2} \begin{bmatrix} 1 & -1 \\ 1 & -1 \end{bmatrix} \qquad [3.18]$$

The consistent mechanical element load vector due to thermal expansion in Eq. [3.11] is

$$\boldsymbol{f}_{\mathrm{m}} = \int_{v^{(e)}} \boldsymbol{N}_{\mathrm{m}}^{T}\left(-E\alpha\frac{\mathrm{d}T}{\mathrm{d}x}\right)\mathrm{d}v = -\int_{v^{(e)}} \boldsymbol{N}_{\mathrm{m}}^{T} E\alpha \boldsymbol{B}_{\mathrm{th}} \mathrm{d}v \boldsymbol{T}^{(e)} \qquad [3.19]$$

where $\boldsymbol{B}_{\mathrm{th}}$ gives the temperature gradient in the element from the nodal temperatures. It contains the derivatives of the shape functions used for interpolating the displacements in the element. The coupling matrix for a two-node element becomes

$$\boldsymbol{m}_{\text{th}}^{(\text{e})} = \frac{EA\alpha}{2}\begin{bmatrix} 1 & -1 \\ 1 & -1 \end{bmatrix} \tag{3.20}$$

Note that $\boldsymbol{m}_{\text{u}}$ and $\boldsymbol{m}_{\text{th}}$ are equivalent except for the T_{ref} term when the same shape functions are used for the mechanical and thermal fields. This has not been utilised in the following equations, as in Armero and Simo [23].

The standard procedure of assembling element vectors and matrices leads to a fully coupled, asymmetric, system of equations for the thermoelastic problem

$$\begin{bmatrix} \boldsymbol{M} & 0 \\ 0 & 0 \end{bmatrix}\begin{bmatrix} \ddot{\boldsymbol{U}} \\ \ddot{T} \end{bmatrix} + \begin{bmatrix} 0 & 0 \\ \boldsymbol{M}_{\text{u}} & \boldsymbol{C} \end{bmatrix}\begin{bmatrix} \dot{\boldsymbol{U}} \\ \dot{T} \end{bmatrix} + \begin{bmatrix} \boldsymbol{K} & \boldsymbol{M}_{\text{th}} \\ 0 & \boldsymbol{K}_{\text{th}} \end{bmatrix}\begin{bmatrix} \boldsymbol{U} \\ T \end{bmatrix} = \begin{bmatrix} 0 \\ 0 \end{bmatrix} \tag{3.21}$$

where the submatrix $\boldsymbol{M}$ is obtained from assembling element mass matrices, $\boldsymbol{K}$ from element stiffness matrices, $\boldsymbol{C}$ from element heat capacity matrices and $\boldsymbol{K}_{\text{th}}$ from element conductivity matrices. The coupling matrices $\boldsymbol{M}_{\mathbf{th}}$ and $\boldsymbol{M}_{\mathbf{u}}$ are obtained from assembling the above described element coupling matrices.

The temporal discretisation is performed using the fully implicit version of Newmark's method for the mechanical part [33, 35]. This is

$${}^{n+1}\boldsymbol{U} = {}^{n}\boldsymbol{U} + \Delta t\,{}^{n}\dot{\boldsymbol{U}} + \frac{\Delta t^2}{2}\,{}^{n+1}\ddot{\boldsymbol{U}} \tag{3.22}$$

and

$${}^{n+1}\dot{\boldsymbol{U}} = {}^{n}\dot{\boldsymbol{U}} + \Delta t\,{}^{n+1}\ddot{\boldsymbol{U}} \tag{3.23}$$

They can be re-formed to the updated formulas:

$${}^{n+1}\ddot{\boldsymbol{U}} = \frac{2}{\Delta t^2}\left({}^{n+1}\boldsymbol{U} - {}^{n}\boldsymbol{U}\right) - \frac{2}{\Delta t}\,{}^{n}\dot{\boldsymbol{U}} \tag{3.24}$$

and

$${}^{n+1}\dot{\boldsymbol{U}} = \frac{2}{\Delta t}\left({}^{n+1}\boldsymbol{U} - {}^{n}\boldsymbol{U}\right) - {}^{n}\dot{\boldsymbol{U}} \tag{3.25}$$

An implicit time stepping algorithm is also used for the rate of change in temperature [24, 33]:

$${}^{n+1}\boldsymbol{T} = {}^{n}\boldsymbol{T} + \Delta t\,{}^{n+1}\dot{\boldsymbol{T}} \rightarrow {}^{n+1}\dot{\boldsymbol{T}} = \frac{{}^{n+1}\boldsymbol{T} - {}^{n}\boldsymbol{T}}{\Delta t} = \frac{\Delta \boldsymbol{T}}{\Delta t} \tag{3.26}$$

The above relations can be rewritten to give acceleration, velocities and temperature rates at end of a time step, ${}^{n+1}t$. They are inserted into Eq. [3.21], giving

$$\begin{bmatrix} K + \dfrac{2}{\Delta t^2} M & M_{\text{th}} \\ \dfrac{2}{\Delta t} M_{\text{u}} & \dfrac{1}{\Delta t} C + K_{\text{th}} \end{bmatrix}^{n+1} \begin{bmatrix} U \\ T \end{bmatrix} =$$

$$\begin{bmatrix} \dfrac{2}{\Delta t} M \\ M_{\text{u}} \end{bmatrix} {}^{n}\dot{U} + \begin{bmatrix} \dfrac{2}{\Delta t^2} M & 0 \\ \dfrac{2}{\Delta t} M_{\text{u}} & \dfrac{1}{\Delta t} C \end{bmatrix}^{n} \begin{bmatrix} U \\ T \end{bmatrix} \tag{3.27}$$

The solution of the fully coupled problem above is possible but seldom efficient. Different staggered approaches to solve this coupled problem are described below and applied in the example in Box 3.3.

3.2.3 Staggered solution starting with deformation problem

A staggered solution of the problem can be formulated in different ways. One possibility is to compute the deformation first using the temperatures at ${}^{n}t$ as predictor values. The semi-discrete problem then becomes

$$M^{n+1}\ddot{U} + K^{n+1}U = -M_{\text{th}}{}^{n}T \tag{3.28}$$

The time stepping logic in Eqs [3.24] and [3.25] gives

$$\left(K + \frac{2}{\Delta t^2} M\right)^{n+1} U = \frac{2}{\Delta t} M^{n}\dot{U} + \frac{2}{\Delta t^2} M^{n}U - M_{\text{th}}{}^{n}T \tag{3.29}$$

The thermal analysis is performed thereafter by solving

$$C^{n+1}\dot{T} + K_{\text{th}}{}^{n+1}T = -M_{\text{u}}{}^{n+1}\dot{U} \tag{3.30}$$

with the fully implicit time stepping logic in Eq. [3.26] giving

$$\left(\frac{1}{\Delta t} C + K_{\text{th}}\right)^{n+1} T = \frac{1}{\Delta t} C^{n}T - M_{\text{u}}{}^{n+1}\dot{U} \tag{3.31}$$

3.2.4 Staggered solution starting with thermal problem

Another variant of the staggered solution approach is to solve the temperature field first using a predictor for the velocity. This is expressed as

$$C^{n+1}\dot{T} + K_{\text{th}}{}^{n+1}T = -M_{\text{u}}{}^{n}\dot{U} \tag{3.32}$$

The use of the fully implicit time stepping logic in Eq. [3.26] gives

$$\left(\frac{1}{\Delta t} C + K_{\text{th}}\right)^{n+1} T = \frac{1}{\Delta t} C^{n}T - M_{\text{u}}{}^{n}\dot{U} \tag{3.33}$$

Thereafter the mechanical field is advanced using the computed temperatures by solving

$$M^{n+1}\ddot{U} + K^{n+1}U = -M_{\text{th}}\,{}^{n+1}T \qquad [3.34]$$

The time stepping logic in Eqs [3.24] and [3.25] gives

$$\left(K + \frac{2}{\Delta t^2}M\right){}^{n+1}U = \frac{2}{\Delta t}M^n\dot{U} + \frac{2}{\Delta t^2}M^nU - M_{\text{th}}\,{}^{n+1}T \qquad [3.35]$$

3.2.5 Staggered solution starting with augmented thermal problem

Farhat and coworkers [20] proposed an unconditional stable approach, called an augmented staggered procedure. They use an improved predictor for the velocity when advancing the temperature solution in Eq. [3.33]. This is based on some of the relations from the mechanical solution which is what they define as the augmentation procedure. The velocity is predicted, consistent with Eq. [3.23], to be

$${}^{\mathrm{P}}\dot{U} = {}^{n}\dot{U} + \Delta t^{n+1}\ddot{U} \qquad [3.36]$$

The equation of motion, Eq. [3.34], is inserted above in order to give the acceleration at the end of the increment. This gives

$${}^{\mathrm{P}}\dot{U} = {}^{n}\dot{U} - \Delta t M^{-1}\left(K^{n+1}U + M_{\text{th}}\,{}^{n+1}T\right) \qquad [3.37]$$

The displacements ${}^{n+1}U$ are approximated by ${}^{n}U$ and this gives the improved predictor for the velocity

$${}^{\mathrm{P}}\dot{U} = {}^{n}\dot{U} - \Delta t M^{-1}\left(K^nU + M_{\text{th}}\,{}^{n+1}T\right) \qquad [3.38]$$

This is inserted into Eq. [3.33] with ${}^{P}\dot{U}$ replacing ${}^{n}\dot{U}$ in the last term on the right hand side. Some manipulations give the augmented heat conduction equation

$$\left(\frac{1}{\Delta t}C + K_{\text{th}} - \Delta t M_{\text{u}} M^{-1} M_{\text{th}}\right){}^{n+1}T$$

$$= \frac{1}{\Delta t}C^nT - M_{\text{u}}\,{}^{n}\dot{U} + \Delta t M_{\text{u}} M^{-1} K^n U \qquad [3.39]$$

which is solved first. The matrix on the left hand side is symmetric provided the same interpolation functions are used for the thermal and mechanical fields. Thereafter the mechanical part of the problem, Eq. [3.35], is solved.

3.2.6 Staggered solution based on adiabatic split

Armero and Simo [23] proposed a similar unconditional stable procedure as Farhat and coworkers [20] but starting with the mechanical step. They use the notations of operator split when discussing the solution of the coupled problems.

Here, I make a somewhat different presentation of their algorithm in order to elucidate its relation to the approach proposed by Farhat *et al.* The implementation in Armero and Simo is more efficient as they use a lumped heat capacity matrix enabling an analytic update. They also utilise $\boldsymbol{m}_{\mathrm{u}} = T_{\mathrm{ref}}\boldsymbol{m}_{\mathrm{th}}$ since the same shape functions are used for the mechanical and thermal fields. The mechanical stiffness matrix is obtained from the standard stiffness matrix using adiabatic elastic properties (see Box 3.2). Armero and Simo [23] call the staggered approach according to Eqs [3.29] and [3.31] an isothermal split. They derive a stability criterion for this approach.

$$\Delta t \leq \frac{L_{\mathrm{min}}}{\sqrt{E/\rho}} \frac{2}{E\alpha\sqrt{\dfrac{T_{\mathrm{ref}}}{E\rho c_{\mathrm{v}}}}} = \Delta t_{\mathrm{wave}} \frac{2}{\sqrt{\dfrac{{}_{\mathrm{ad}}E}{E} - 1}} \tag{3.40}$$

where Δt_{wave} is the time for an acoustic wave to pass through the smallest element in the model of length $L_{\mathrm{min}} \cdot \sqrt{E/\rho}$ is the speed of this wave, i.e. the speed of sound in the material and ${}_{\mathrm{ad}}E$ is the adiabatic Young's modulus. The difference between the adiabatic modulus and elastic modulus is very small for metals.

The unconditional stable procedure proposed by them is called an adiabatic, or isentropic, split. This corresponds to an augmentation of the mechanical analysis using an improved predictor of the temperatures when advancing the mechanical solution. They replace the temperature ${}^{n}T$ in Eq. [3.28] with an improved estimate by accounting for adiabatic heating. This predictor ${}^{\mathrm{P}}T$ is obtained by removing the heat conduction part in Eq. [3.30]. This is the second term on the left hand side in the equation. Then the improved predictor for the temperature rate when advancing the mechanical analysis is computed as

$${}^{\mathrm{P}}\dot{\boldsymbol{T}} = -\boldsymbol{C}^{-1}\boldsymbol{M}_{\mathrm{u}}\,{}^{n+1}\dot{\boldsymbol{U}} \tag{3.41}$$

Thereafter the finite difference approximation of the temperature rate from Eq. [3.26] gives the wanted predictor

$${}^{\mathrm{P}}\boldsymbol{T} = {}^{n}\boldsymbol{T} - \Delta t\boldsymbol{C}^{-1}\boldsymbol{M}_{\mathrm{u}}\,{}^{n+1}\dot{\boldsymbol{U}} \tag{3.42}$$

This is inserted into the last term in the equation of motion, Eq. [3.28]. The time stepping formulas, Eqs [3.24] and [3.25], are then applied, giving

$$\begin{aligned}&\left(\boldsymbol{K} - 2\boldsymbol{M}_{\mathrm{th}}\boldsymbol{C}^{-1}\boldsymbol{M}_{\mathrm{u}} + \frac{2}{\Delta t^2}\boldsymbol{M}\right){}^{n+1}\boldsymbol{U}\\ &= \frac{2}{\Delta t^2}\boldsymbol{M}\left(\Delta t^{n}\dot{\boldsymbol{U}} + {}^{n}\boldsymbol{U}\right) - \boldsymbol{M}_{\mathrm{th}}\boldsymbol{C}^{-1}\boldsymbol{M}_{\mathrm{u}}\left(\Delta t^{n}\dot{\boldsymbol{U}} + 2^{n}\boldsymbol{U}\right) - \boldsymbol{M}_{\mathrm{th}}\,{}^{n}\boldsymbol{T}\end{aligned} \tag{3.43}$$

which is solved. Thereafter the heat conduction problem is solved without the mechanically generated heat but the predicted temperatures ${}^{\mathrm{P}}\boldsymbol{T}$ as the initial state instead of ${}^{n}\boldsymbol{T}$:

$$\left(\frac{1}{\Delta t}\boldsymbol{C}+\boldsymbol{K}_{\mathrm{th}}\right)^{n+1}\boldsymbol{T}=\frac{1}{\Delta t}\boldsymbol{C}\boldsymbol{T}^{\mathrm{P}}$$

$$\text{or}\quad\left(\frac{1}{\Delta t}\boldsymbol{C}+\boldsymbol{K}_{\mathrm{th}}\right)^{n+1}\boldsymbol{T}=\frac{1}{\Delta t}\boldsymbol{C}^{n}\boldsymbol{T}-\boldsymbol{M}_{\mathrm{u}}{}^{n+1}\dot{\boldsymbol{U}} \qquad [3.44]$$

3.2.7 Application of solution methods on a vibrating bar

Five different approaches to solve coupled thermoelastic problems are shown above. The relations are general provided appropriate matrices are inserted, although we started with the one-dimensional field equations in Section 3.2.1. The methods are illustrated in Box 3.3 for the one-dimensional example taken from Armero and Simo [23]. The results presented are based on a time step just above the stability limit of the classical staggered approaches. One can see that they become unstable whereas the augmented procedures show no such tendency.

Box 3.3 Solutions for one-dimensional thermoelastic problem

A bar with fixed ends held at zero temperature is analysed [23] (see Fig. 3.5). It has an initial temperature 300 K at which is stress free. This is also taken as the reference temperature in Eq. [3.11]. The temperature change from this temperature is computed. The initial load is the initial velocity of the bar. It is assumed to be

$$\dot{u}_0=\sin\left(\frac{\pi x}{L}\right) \qquad [3.45]$$

The governing equations are given in Section 3.2.1 and the numerical solution methods in Sections 3.2.2–6 are used. The used data can be found in the available code and are not the same as in [23]. An analysis using a time step that is 0.6% higher than the stability limit of the isothermal split is used. The computed temperature is shown in Fig. 3.6 and computed displacement in Fig. 3.7. FC in the figures denotes the results obtained from the fully coupled solution procedure in Section 3.2.2. M/T denotes the

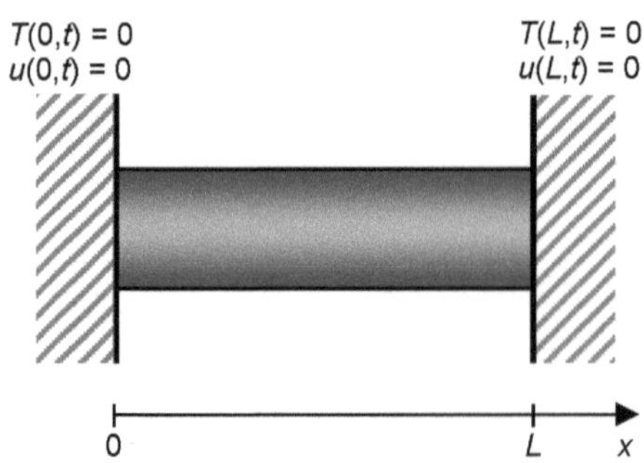

3.5 Bar with axial vibrations. Boundary conditions are indicated in the figure.

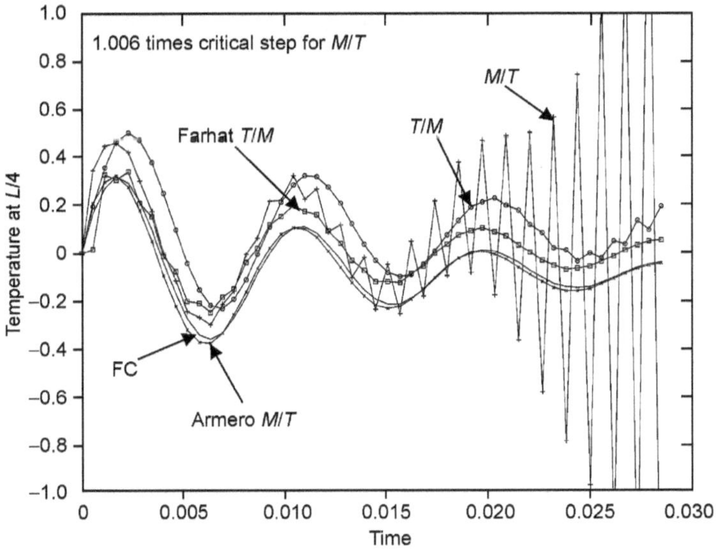

3.6 Temperature history at $x = L/4$.

classical split with an isothermal mechanical step followed by a thermal step. T/M is the other way around. It is possible to see the instability of the classical splits. The method proposed by Armero and Simo [23] is more accurate than the one given by Farhat and coworkers [20].

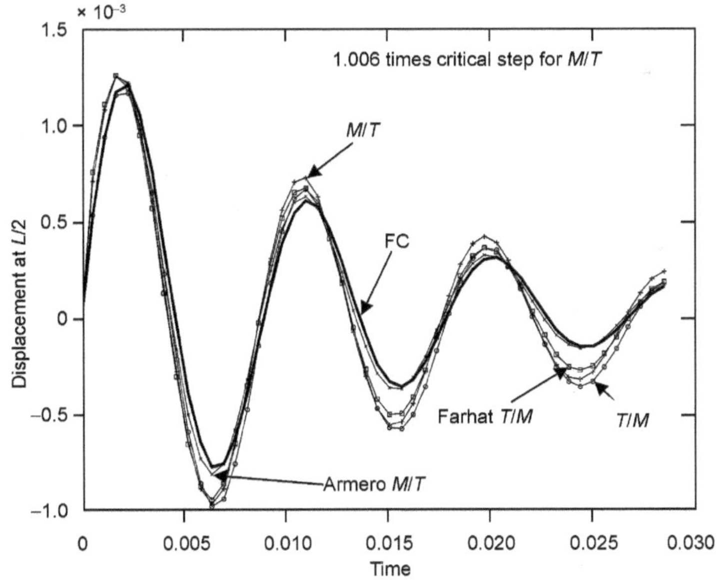

3.7 Displacement history at $x = L/2$.

3.3 Decoupling of the subdomains of welding simulations

A general view of the relevant fields in welding simulations is shown in Fig. 3.8 where the box 'physics of heat generation' is a generic representation for all possible welding processes. Simulations of this type require weld process models in combination with the CWM model. An example of this kind of multi-physics was introduced in Chapter 2 and is not discussed any further in this book. The distribution of heat input is usually predefined in CWM models. It is determined by calibrating the model with respect to measurements as described in Section 8.5.2. A heat input model as illustrated in Fig. 3.9 replaces the 'physics of heat generation' box. The work by Chen and Sheng [18, 19] is of this type. The domains of the fields 'fluid flow' and 'deformation in solid' in Fig. 3.8 have a common interface at the weld pool boundary. This is a type of fluid–structure interaction (FSI) [20] where the boundary between them is not known beforehand.

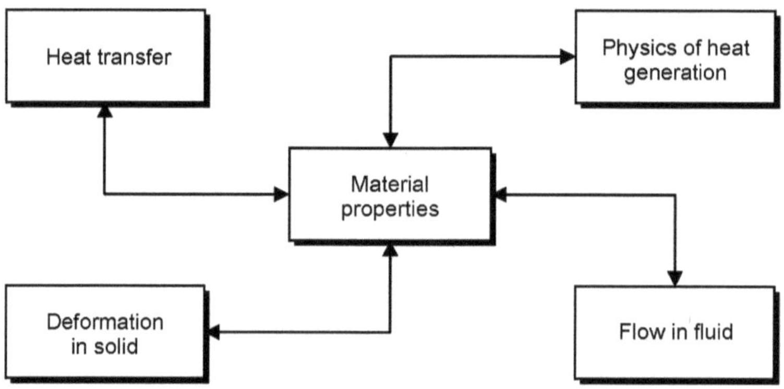

3.8 Different field equations in CWM together with weld process models.

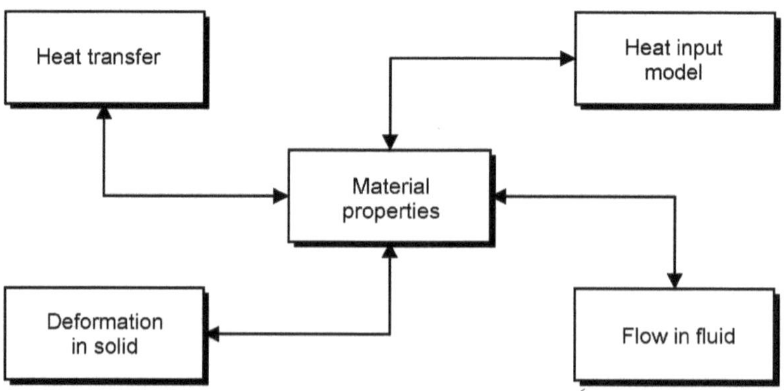

3.9 Fields in CWM modelling of fusion welding without a weld process model but with a fluid flow model.

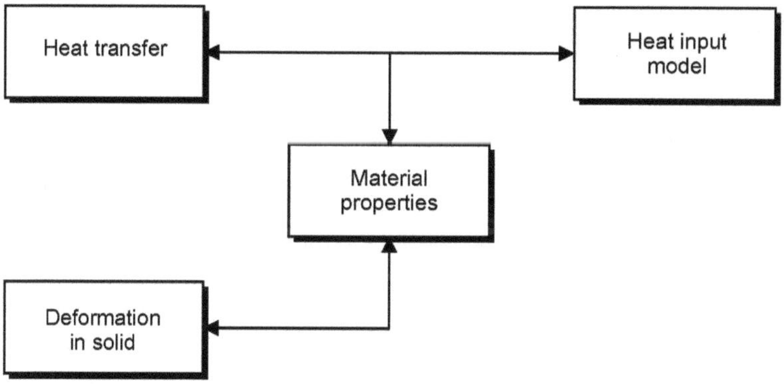

3.10 Fields in classical CWM modelling of fusion welding without a welding process model and without fluid flow.

Most analyses in CWM ignore the fluid flow and prescribe the distribution of the heat input, and the coupling scheme then becomes as shown in Fig. 3.10. A fully coupled solution of these fields was used by Wang *et al.* [36]. Ronda and Oliver [37] even included phase composition into the total system of equations. Finite element formulation of coupled thermoelastoplastic problems is given in Hsu [25] and Sluzalec [27]. The unconditional stable adiabatic split proposed by Armero and Simo [23, 38] and introduced in Section 3.2.6 has not, so far, been used by anyone in CWM simulations. This is probably because, although it can be efficiently implemented, it requires code modifications compared with the standard implementations of mechanical and thermal analysis codes. An adiabatic update of the temperature in the mechanical phase must be introduced. Furthermore, the stress update must be done with respect to the yield surface for the temperature at beginning of the increment. Thereafter, a further stress update under frozen deformation must be done in the thermal phase where the yield surface for the final temperature is used. The stability limit of the time stepping in the traditional isothermal split has not been found to cause any problem in CWM when using a staggered approach accounting for mechanically generated heat. This probably because the coupling term is small and the demand for accuracy limits the time steps used. Naturally, stability is no problem in solutions based on the staggered approach when the mechanically generated heat is excluded from the equations. The approach proposed by Armero and Simo [23, 38] may be an efficient alternative when solving problems where the thermomechanical coupling is strong, as in explosive welding. The most common way is to use the monolithic approach as discussed in Section 3.2.2 to solve the coupled thermomechanical problem in Fig. 3.10.

The coupling between material behaviour and temperature and deformation fields is shown in Fig. 3.11 and explained in Table 3.1. The plastic dissipated energy, coupling no. 2a, is the largest contribution to the mechanically generated heat but still negligible compared with the heat input [39]. Furthermore, if the

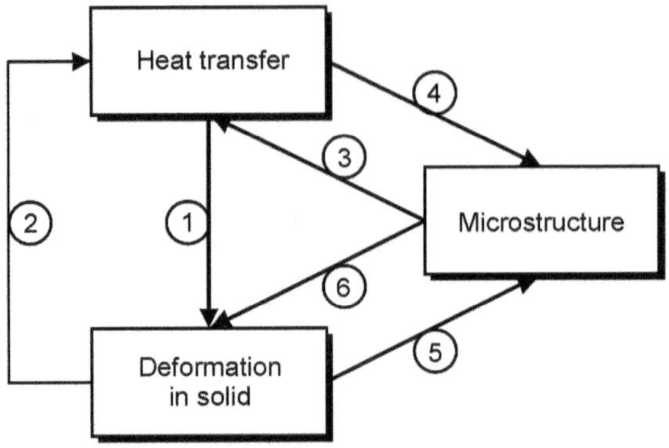

3.11 Couplings in thermomechanical models.

Table 3.1 Thermomechanical couplings in Fig. 3.11

Coupling	Description
1	Temperature changes drive the deformation via thermal expansion and volume changes due to phase changes denoted by coupling no 6.
2	(a) Deformation generated heat. (b) Deformation affects thermal boundary conditions.
3	Thermal properties depend on microstructure and phase changes are associated with latent heats.
4	Thermal driven phase changes.
5	Deformation driven phase changes.
6	The mechanical material behaviour depends on the microstructure and temperature.

effect of the deformation on thermal boundary conditions, coupling no. 2b, can be ignored, then a weakly coupled analysis can be done. Then the complete simulation of the heat flow is followed by the deformation simulation. The temperature is read from the file in the latter simulation. This file was saved during the thermal simulation. However, it is very convenient to use a staggered approach, Fig. 3.1, for weakly coupled problems as the thermal analysis does not add much to the required computer time and one does not need the book-keeping necessary for ensuring that the correct temperature file is read during a subsequent mechanical analysis. Figure 3.12 shows the staggered procedure, which is convenient to use in CWM simulations. Ronda and Oliver [37] named the fully coupled problem as a thermo-metallurgical-mechanical (TMM) problem. However, the staggered approach can also be used to solve TMM problems provided the appropriate models for the microstructure evolution are implemented, see Section 9.9. More details about heat input models are given in Section 9.12.

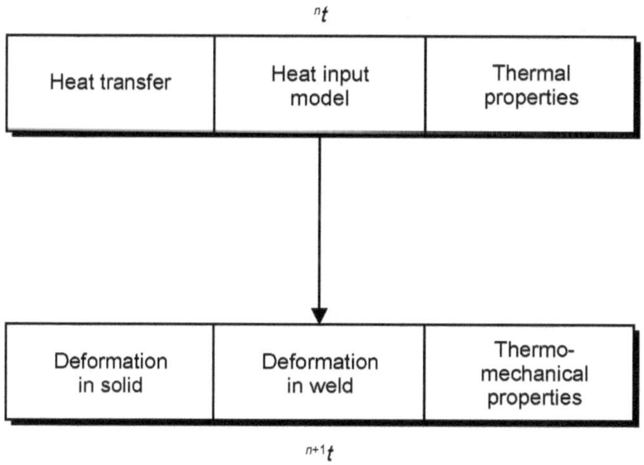

3.12 Staggered approach using isothermal split starting with a heat conduction analysis with fixed geometry.

Thermal stress problems can usually be treated as quasi-static problems, in which case the inertia forces are ignored in the mechanical analysis. This is also the case for welding processes, with the exception of explosive welding where the deformation generates the heat.

Two basic variants of reference frames have been used in CWM. The Lagrangian approach is the most common one but Eulerian frames have also been used. The latter can take advantage of the steady state conditions that may exist with respect to the moving heat source. Eulerian analyses have been published in references [40–50]. The two different types of reference frames are shown in Fig. 3.13 in the case of butt welding of a plate. They are also discussed in the book by Goldak and Akhlaghi [51]. Computational savings

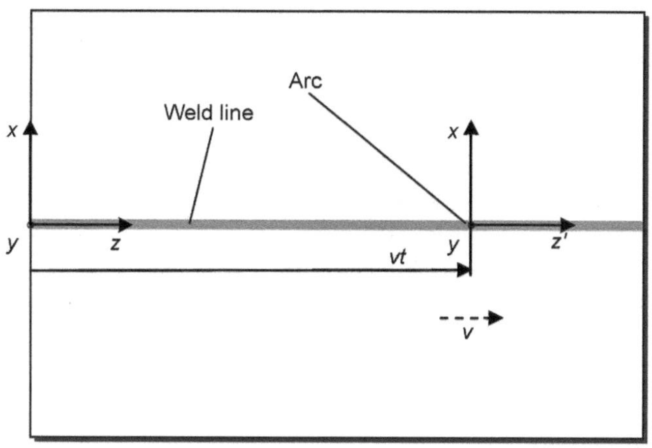

3.13 Lagrangian and co-moving reference frames.

from one to three orders of magnitude as compared with Lagrangian analyses are reported.

Co-moving coordinate systems for welding have long been applied: the analytic solutions by Rosenthal [52, 53] for the temperature fields due to moving point and line sources used this approach. The heat conduction equation in an Eulerian reference frame transforms to

$$\rho c(\dot{T} + \boldsymbol{v} \cdot \nabla T) = \nabla \cdot (\lambda \nabla T) + \dot{Q} \qquad [3.46]$$

where c is the heat capacity, $\boldsymbol{v}$ is the velocity vector of the mass flow with respect to the coordinate system and $\dot{Q}$ is the heat generation per unit volume. Thus, the heat source is moving in the direction of $-\boldsymbol{v}$. This velocity vector is the same as the velocity of the heat source. The assumption that steady state conditions prevail with respect to the moving heat source leads to

$$\rho c \boldsymbol{v} \cdot \nabla T - \nabla \cdot (\lambda \nabla T) = \dot{Q} \qquad [3.47]$$

The equation of motion can be transformed in the same manner.

These equations are the start of the finite element formulation and lead to an asymmetric system of equations due to the convective term. Furthermore, the standard finite element formulation, the Galerkin formulation, is unstable [34, 54].[1] The stability is remedied, and accuracy improved, by enhancing the conductivity. This is illustrated for the one-dimensional problem described in Box 3.4. This enhancement must be done along the streamlines in order to avoid crosswind diffusion when implemented in two- or three-dimensional formulations. This is called the stream-upwind-Petrov–Galerkin (SUPG) method [55]. The diffusivity enhancement discussed in Box 3.4 can be derived by using test functions that emphasise the conduction in the flow direction, which motivates the name Petrov–Galerkin methods. Furthermore, the material response in the elastoplastic analysis must be found by integrating along the streamlines in order to include the history effect. However, the strength of the Eulerian formulation is its effectiveness, which makes it possible to have a fine three-dimensional mesh near the moving arc. The assumption of steady state conditions with respect to the arc limits the applicability of this formulation. For example, the start and finish of the butt welding of a plate cannot be studied.

The generality and availability of Lagrangian codes make them a common choice for CWM models. Belytschko *et al.* [34] described the updated (UL) and total (TL) Lagrangian approaches. The latter is the choice in CWM and therefore the finite element formulations for heat flow and deformation, given in Chapters 5 and 6, are given only for the UL Lagrangian approach.

1. Most papers use the notation diffusivity as they discuss convection–diffusivity problems. The notation for diffusivity has been replaced by conductivity here as we discuss heat conduction problems.

Box 3.4 Combined convective and conductive heat transfer

The heat transfer in a material moving with the velocity v in positive direction and no volume heats is governed by

$$\rho c v \frac{dT}{dx} - \lambda \frac{d^2T}{dx^2} = 0 \quad [3.48]$$

The analytic solution of this problem with the boundary conditions

$$T(0) = T_{\text{left}} \quad [3.49]$$

$$T(L) = T_{\text{right}} \quad [3.50]$$

is

$$T(x) = T_{\text{left}} + \left(T_{\text{right}} - T_{\text{left}}\right) \frac{\left(e^{2Pe^{*}x/L^{(e)}} - 1\right)}{\left(e^{2Pe^{*}L/L^{(e)}} - 1\right)} \quad [3.51]$$

where

$$Pe = \frac{\rho c v L^{(e)}}{2\lambda} \quad [3.52]$$

is the Peclet number. It measures the relative strengths of convection and conduction of heat in an element.

The finite element solution leads to a global system of equations

$$\boldsymbol{K}_{\text{th}} \boldsymbol{T} = \boldsymbol{0} \quad [3.53]$$

with standard contributions for conduction part using a two-node linear element is

$$\boldsymbol{k}_{\text{th}}^{(e)} = \int_{v^{(e)}} \boldsymbol{B}_{\text{th}}^{T} \lambda \boldsymbol{B}_{\text{th}} dv = \frac{\lambda}{L^{(e)}} \begin{bmatrix} 1 & -1 \\ -1 & 1 \end{bmatrix} \quad [3.54]$$

The additional contribution due to the convection part is [34, 54, 56, 57]

$$\boldsymbol{k}_{\text{thconv}}^{(e)} = \int_{v^{(e)}} \boldsymbol{N}_{\text{th}}^{T} \lambda \boldsymbol{B}_{\text{th}} dv = \frac{\rho c v}{2} \begin{bmatrix} -1 & 1 \\ -1 & 1 \end{bmatrix} \quad [3.55]$$

Summing the two matrices above gives the element matrix

$$\boldsymbol{k}_{\text{th}}^{(e)} = \frac{\lambda}{L^{(e)}} \left[\begin{bmatrix} 1 & -1 \\ -1 & 1 \end{bmatrix} + Pe \begin{bmatrix} -1 & 1 \\ -1 & 1 \end{bmatrix} \right] = \frac{\lambda}{L^{(e)}} \begin{bmatrix} 1 - Pe & -1 + Pe \\ -1 - Pe & 1 + Pe \end{bmatrix} \quad [3.56]$$

to be assembled into Eq. [3.53]. This formulation becomes unstable when $Pe = 1$ [34] as can be seen in Fig. 3.14.

Equation [3.56] becomes

$$\boldsymbol{k}_{\text{th}}^{(\text{e})} = \frac{\lambda}{L^{(\text{e})}} \begin{bmatrix} 1 - Pe & -1 + Pe \\ -1 - Pe & 1 + Pe \end{bmatrix} \quad [3.57]$$

This conditional stable formula is stabilised by enhancing the conductivity so that the Peclet number is modified to ${}_{\text{mod}}Pe$ according to the equation below.

$${}_{\text{mod}}Pe = \tanh(Pe) \quad [3.58]$$

This value is inserted into Eq. [3.58] instead of Pe in the stabilised formulation.

The classical formulation is compared with the stabilised formulation and theoretical solutions for $Pe = 1$ and 3 in Figs 3.14 and 3.15. The improved formulation is a prelude to SUPG-formulations.

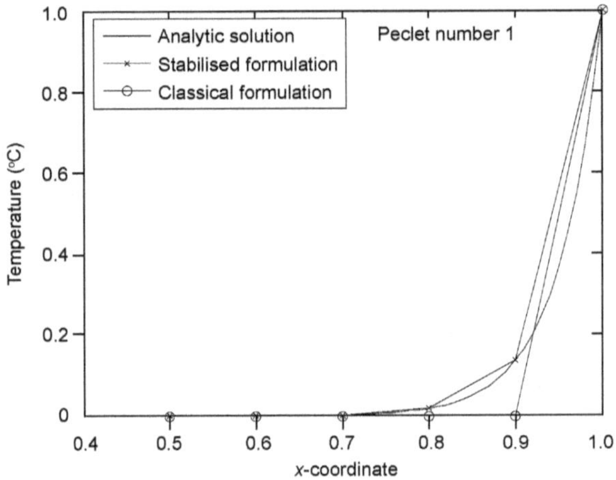

3.14 Temperature in fluid flow with given inlet and outlet temperatures for $Pe = 1$.

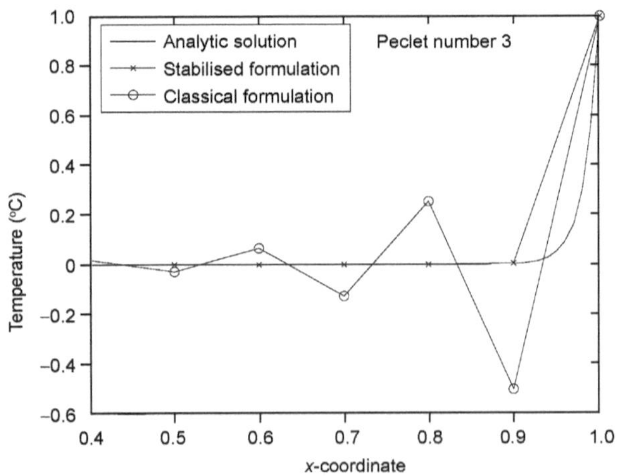

3.15 Temperature in fluid flow with given inlet and outlet temperatures for $Pe = 3$.

Box 3.5 Moving point source

The heat transfer in a material moving with the velocity v in positive direction and a point source at the coordinate x_c is

$$\rho c v \frac{dT}{dx} - \lambda \frac{d^2T}{dx^2} = P\delta(x - x_c) \qquad [3.59]$$

where v is now interpreted as the material flow relative to the heat source. Thus the heat source is moving towards the left in Figs 3.16 and 3.17. The boundary conditions and the finite element formulations are the same as in Box 3.4.

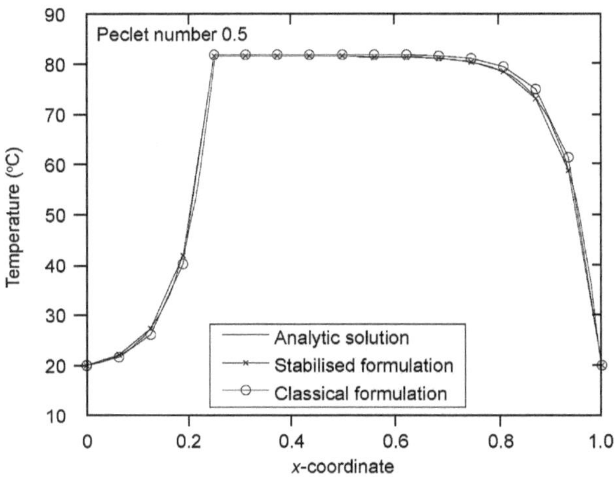

3.16 Temperature due to a point heat source moving to the left, $Pe = 0.5$.

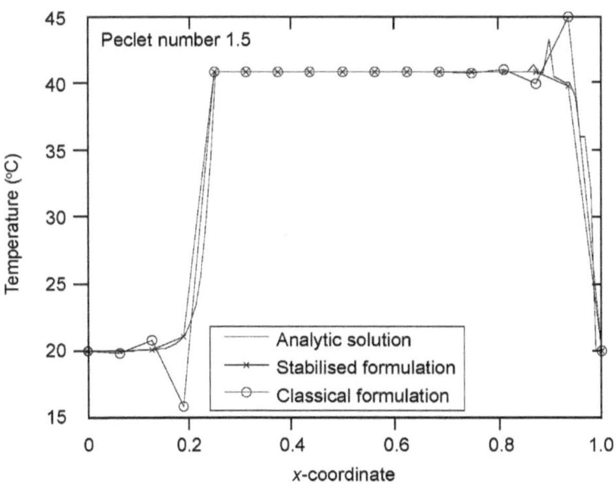

3.17 Temperature due to a point heat source moving to the left, $Pe = 1.5$. Wiggles at end of analytic solution due to inaccuracy in its evaluation.

It is assumed that the point source is applied on a node. Then there will be one contribution to the load vector and Eq. [3.53] becomes

$$\boldsymbol{K}_{\mathrm{th}}\boldsymbol{T} = \dot{\boldsymbol{Q}}_{\mathrm{ext}} \qquad [3.60]$$

The enhancement of the heat conductivity that gives $_{\mathrm{mod}}Pe$ in Eq. [3.58] is

$$_{\mathrm{mod}}\lambda = \lambda(1 + \alpha Pe) \qquad [3.61]$$

where

$$\alpha = \coth(Pe) - \frac{1}{Pe} \qquad [3.62]$$

The value $P/(1 + \alpha Pe)$ is placed at the appropriate position in $\dot{\boldsymbol{Q}}_{\mathrm{ext}}$. The computed temperatures for different Pe numbers are shown in Figs 3.16 and 3.17.

4
Thermomechanics of welding

The purpose of this chapter is to give an introductory understanding of the phenomena that may have to be accounted for in CWM. Details about the physics in the weld pool, such as diffusion of hydrogen or oxygen, are not discussed, since the focus is on CWM models where the weld metal region is represented by a heat input model.

The thermal cycle and associated microstructural changes is the largest challenge for modelling in CWM. An introduction to this and the effect on stresses and deformation is the foundation for understanding the effects of welding and how to model this process. Thus the chapter starts with a discussion of the microstructure evolution and thermal stress development. More details about the material modelling and its relation to microstructure are given in Chapter 9 and Section 12.2. There are a number of books that focus on welding metallurgy. The book by Granjon [58] with illustrative drawings is a good starting point for understanding this subject. Easterling [59] includes equations and discussions of models. A considerable part of the modelling in his book uses the Rosenthal solutions for the thermal field due to a point source [52, 53]. The book by Grøng is more modern and a useful reference about welding metallurgy [60]. The book by Radaj [7] has a wider scope and describes the thermomechanics of welding in a very clear way.

4.1 The thermal cycle and microstructure evolution

The thermal expansion of steels reveals the solid state phase changes and one of their effects on the thermal stresses developed during welding. Measuring the thermal dilatation can also be used to determine phase changes [61]. Figure 4.1 shows the thermal dilatation of a martensitic stainless steel during heating. The initial structure is ferrite (α) and globular carbides ($M_{23}C_6$). The thermal expansion is reduced when the temperature reaches A_{c1}. This is due to the transformation from ferrite to austenite. The transformation is complete at temperature A_{c3}. Thereafter the thermal expansion has a continuous increase of slope due to the ongoing dissolution of carbides in the austenite. This ends at the

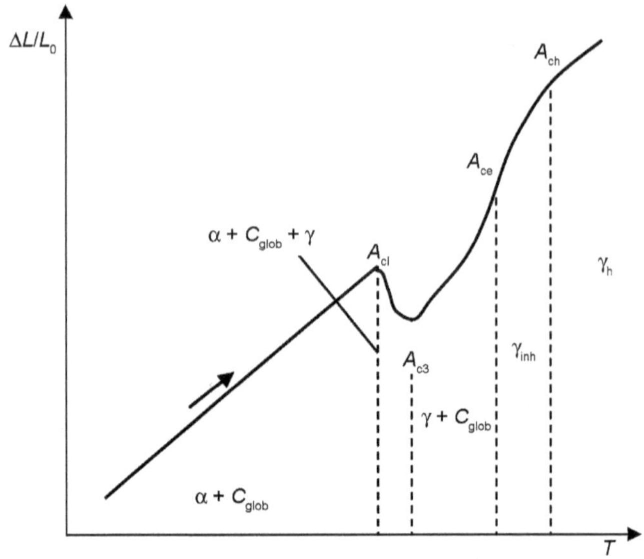

4.1 Dilatation during heating of martensitic stainless steel [61].

inflexion point at the A_{ce} temperature. A_{ch} denotes the temperature at which the dissolved carbon has diffused so that the austenite is relatively homogeneous. The thermal dilatation is the cause of the welding stresses and this figure illustrates its changes during heating. There are naturally changes in elastic and plastic properties of the material due to temperature as well as phase changes.

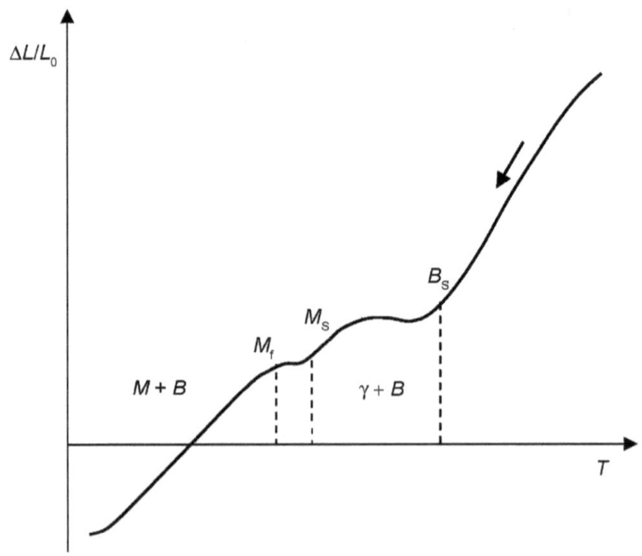

4.2 Dilatation during cooling of a low-carbon manganese steel with 0.07% C, 1.56% Mn and 0.41% Si [61]. The cooling rate was 234 K/s.

The cooling phase is even more important since some of the effects during heating are overshadowed by the last part of the cooling phase in the welding thermal cycle. This is one motivation for the use of cut-off temperature and other similar simplifications used in CWM and discussed in Sections 9.2 and 9.3. A continuous cooling curve is shown in Fig. 4.2. It shows a martensitic transformation in a carbon manganese steel during rapid cooling. The final microstructure is a mixture of martensite and bainite. Bainite starts to form at a higher temperature than martensite and its start formation temperature is denoted by B_s in Fig. 4.2. Thereafter the martensite start, M_s, and finish, M_f, temperatures can be seen.

The temperature changes in welding are very fast and the phase changes depend on the cooling and heating rates and other factors. Some approaches to account for this are given in Section 9.9. However, it is instructive to describe the different microstructure regions near a weld by comparing the peak temperature with the equilibrium Fe–C diagram. Figure 4.3 is a modified version

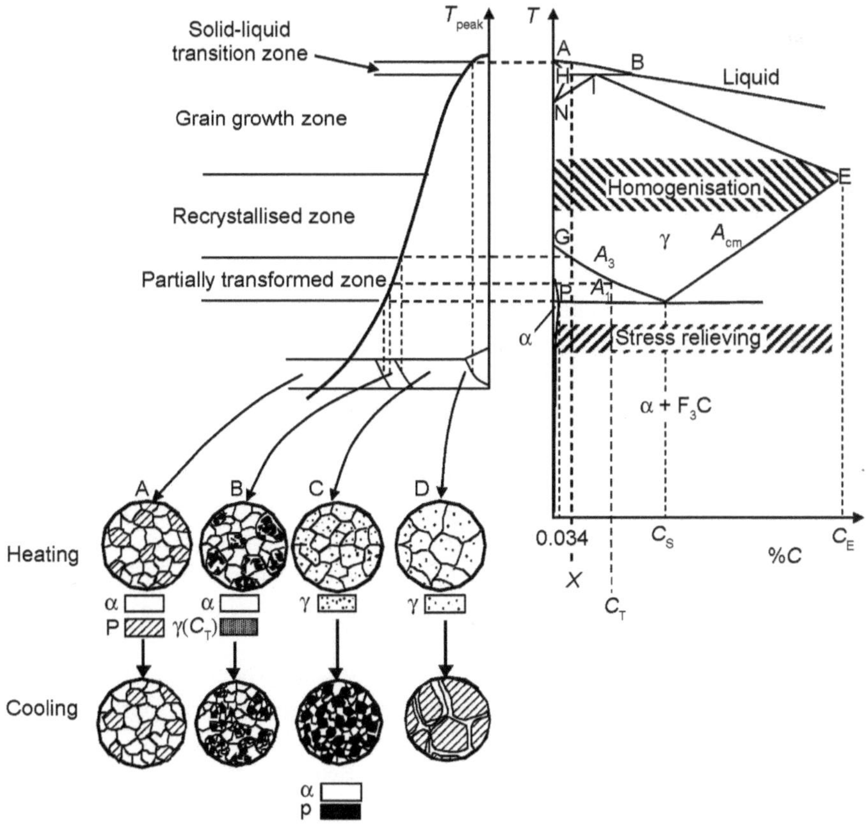

4.3 Thermal cycle and equilibrium Fe–C diagram showing different regions of weld.

of Figure 6.15 in Granjon [58]. The left part of the figure shows the peak temperatures obtained in the weld region. Some different regions are named to the left of the curve. The microstructures after heating to the peak temperatures at four locations (denoted A, B, C and D) are shown below the peak temperature curve. The microstructures after cooling are shown at the bottom of the figure. The equilibrium phase diagram for Fe–C is shown to the right in Fig. 4.3. A steel with carbon fraction less than $C_S = 0.76\%$ is called a hypoeutectoid steel. It is located left of the eutectoid point in the diagram. Steels with a larger fraction of carbon are called hypereutectoid steels. The heat-affected zone (HAZ) is an important concept as it is the region in which the microstructure has changed (regions B–D). The weld region where the material has been molten is denoted the weld metal (WM). The base material (BM) is the region without phase changes. The evolution of the microstructures during heating and subsequent (slow) cooling is described below.

It is assumed that BM is an annealed microstructure with $X\%$ carbon made up of ferrite and pearlite. The latter is a mixture of ferrite and cementite, Fe_3C, with 0.76% C (C_S). The lever rule gives the proportion of ferrite and cementite to match the bulk value of $X\%$ carbon. The temperature rises and the pearlite is transformed directly into austenite at temperature A_1. The temperature rises and ferrite transforms to austenite. The carbon fraction in the austenite then decreases from the eutectoid value of 0.76% towards $X\%$. The transformation is completed when the temperature reaches A_3. Region B is a material that is partially transformed to austenite whereas region C is completely austenised. The difference between regions C and D is that the higher temperature of the latter promotes grain growth. The regions nearest the weld and the weld metal will also be partially or completely molten when the temperature exceeds the lower fusion temperature, $T_{solidus}$. The melting is complete at $T_{liquidus}$.

The regions with austenite upon heating will experience a reverse transformation during cooling. It should be noted that the diagram to the right is an equilibrium diagram. The correct diagrams showing phase changes during heating and cooling must be determined for specific heating and cooling rates. There are a number of different diagrams used to study non-equilibrium phase changes (Chapter 9 in [62]). The HAZ is a small region and often the cooling rate is quite constant; see also Section 9.3. The phase changes during cooling are affected not only by the cooling rate but also by the grain size and carbon content of the transforming austenite.

The microstructures illustrated in Fig. 4.3 would be obtained when the cooling is relatively slow. Then a mixture of ferrite and pearlite will reappear although with different distribution and grain sizes. However, it is more likely that the high cooling rates in welding will create a different microstructure during cooling. This is better illustrated by a continuous cooling diagram (CCT) (see Fig. 4.4). It shows the microstructure that will be obtained when cooling a material from austenitic condition to room temperature with different cooling

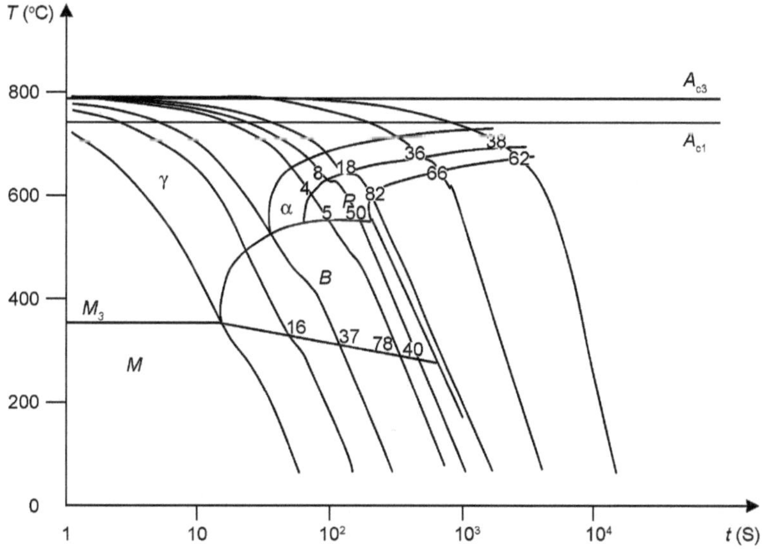

4.4 Continuous cooling diagram for steel with 0.44% C, 0.22% Si, 0.80% Mn, 1.04% Cr, 0.17% Cu and 0.26% Ni (from Chapter 9 in ref. [62]).

rates. The final microstructure will be martensitic if cooled below M_s within less than 10 s. A somewhat slower cooling rate will give a mixture of martensite and bainite. The figures on the cooling curves in the diagram show the volume fraction of microstructure constituents that will remain at room temperature. Special weld CCT diagrams [59] are based on a higher hold temperature before the cooling. This temperature is usually between 1350 and 1400 °C in order to promote grain growth.

Figure 4.5 shows a photo from a T-joint weld showing the different microstructures. Figure 4.6 shows the microstructure along the arrow in Fig. 4.5. It should be read from top-left to bottom-right. The letters denotes the regions in Fig. 4.3. The numerals are used as region delimiters used in the discussion in [63]. The region between 1 and 2 denotes the tempered base metal. The base metal is a mixture of ferrite and pearlite. The large grains developed inside region D′ have been transformed to a fully bainitic-martensitic structure. The difference between this and region D in Fig. 4.3 is that D′ has experienced a phase change. The austenite grains in this region have become so large that the cooling time is not sufficient for the carbon to diffuse from the austenite grain. Then bainite and martensite are formed instead of ferrite-pearlite.

The temper bead technique is a multipass weld technique where subsequent weld passes reheat previous passes. The focus is on creating a better microstructure although this will also affect the stress distribution (see Fig. 4.7). This is discussed in, for example, Radaj [7], Grøng [60] and Leblond *et al.* [64].

The effect of carbide dissolution on the thermal dilatation is visible in Fig. 4.1. Precipitates will grow and also dissolve depending on the temperature. This

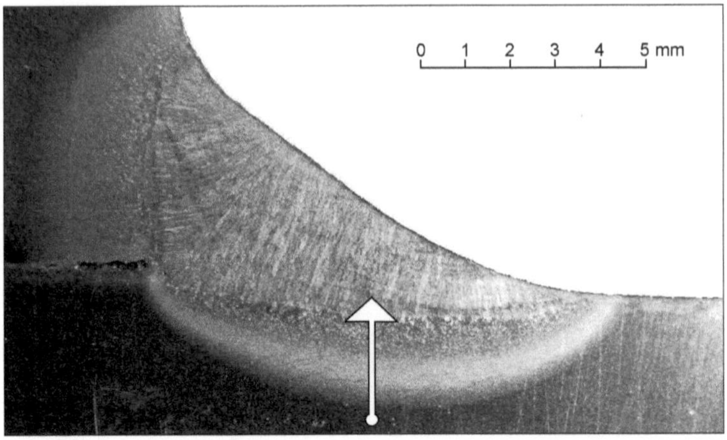

4.5 Photo of T-joint weld. Courtesy L. Fuglsang-Andersen [63].

will affect grain growth and the plastic properties of the material. Easterling [59] and especially Grøng [60] discussed the modelling of coarsening and dissolution of particles. The phenomenon should be appreciated by anyone involved in CWM even if a specific model of the process is seldom used. Diffusion

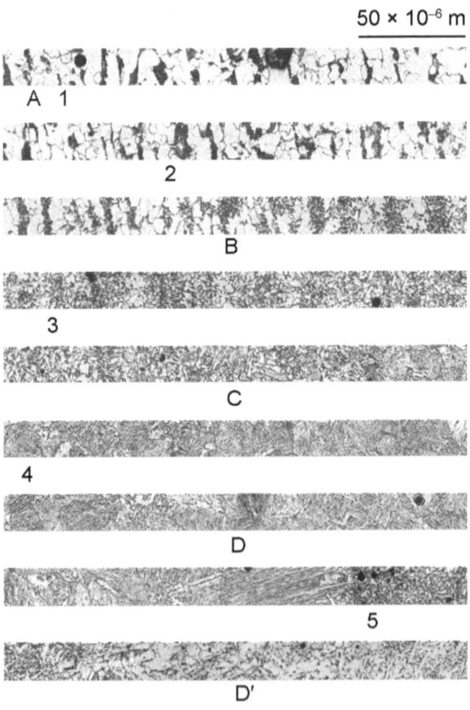

4.6 Microstructure of weld in Fig. 4.5. Letters A–D denotes different regions. Courtesy L. Fuglsang-Andersen [63].

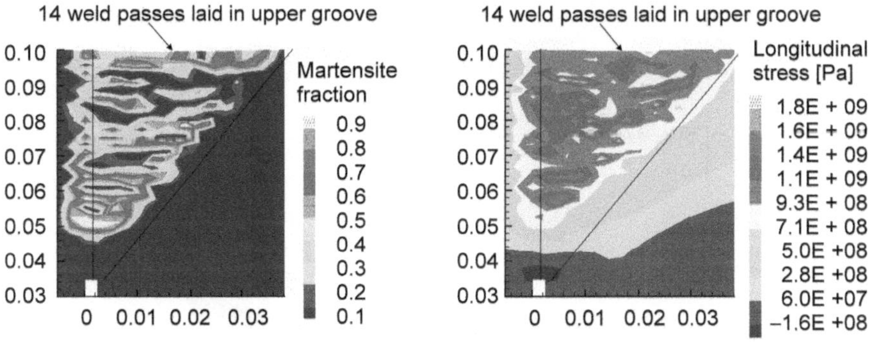

4.7 Effect of sequence of weld passes on martensite fraction and longitudinal residual stress [65]. Fourteen weld passes are laid in the groove.

processes lead to particle growth as a function of time and temperature. However, there is also a solvus temperature at which the particle dissolves. In practice, there is no clear-cut border between these processes as there is a distribution in the size of the particles.

Grain growth is related to the previous discussion of particles. It can be prevented by the presence of precipitates that are pinning the grain boundary. Thus a grain growth model will then only be relevant above the solvus temperature of the precipitates. Additional microstructure features are associated with 'annealing'. This denotes the reduction in hardening due to phase changes, and especially the melting of the material. Then the dislocation density is reduced and the flow stress of the material decreases. The solid state phase changes are also associated with transformation plasticity (TRIP). These two aspects are illustrated in the section below when describing the evolution of stresses during the weld thermal cycles. A discussion of the approaches for simulating a coupled thermo-metallurgical-mechanical model is given in Section 9.9.

4.2 The Satoh test

The evolution of thermal stresses is illustrated by the Satoh test [66]. It is a uniformly heated bar, as shown in Fig. 4.8. The axial stress usually corresponds to the stress in the welding direction. This simplified illustration corresponds to assuming that hot filler is added to the joint and the cold, surrounding material act as a restraint. The heating up, softening and expansion of the surrounding material are ignored in this illustration.

Assuming constant material properties and thermoelastic material behaviour gives the rate of the axial stress

$$\dot{\sigma} = -E_{\mathrm{w}}\alpha_{\mathrm{w}}\dot{T} \qquad [4.1]$$

where E_{w} is Young's modulus and α_{w} is the thermal expansion coefficient of the welded material. The temperature for initial yield is

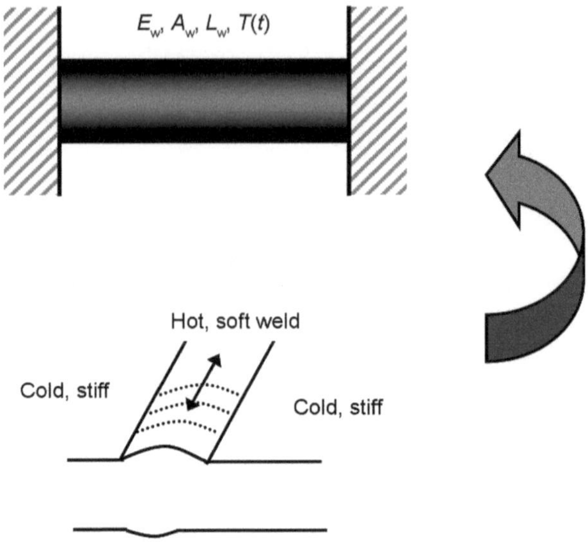

4.8 Satoh test illustrating thermal stress evolution.

$$T^{y} = \frac{E_{w}\alpha_{w}}{\sigma^{y}} = \frac{\alpha_{w}}{\epsilon^{y}} \qquad [4.2]$$

where ϵ^{y} is the strain at plastic yielding and σ^{y} is the yield limit.

Rosenthal's solution for the temperature due to a moving point source in a thin plate [52, 53] can be used to give a relation between peak temperature, T^{p}, at the distance, r, from the weld centre and the weld parameters. This relation is

$$^{p}T - {}^{0}T = \sqrt{\frac{2}{\pi e}}\frac{q}{v}\frac{1}{h}\frac{1}{\rho c}\frac{1}{2r} \qquad [4.3]$$

where ${}^{0}T$ is the initial temperature, q is the power of the heat source per unit length of the weld, h is the plate thickness and v is the welding velocity. Thus q/v is the heat input per unit length [J/m] of the weld. The density is ρ and heat capacity is denoted c. This relation can be found in Easterling [59] and Okerblom (Eq. 5.12) [67]. Ignoring the initial temperature, ${}^{0}T$, and setting the peak temperature equal to T^{y} in Eq. [4.2] gives the width of the plastic region (Okerblom [67], pp. 41–44), as

$$2r^{y} = \sqrt{\frac{2}{\pi e}}\frac{q}{v}\frac{1}{h}\frac{1}{\rho c}\frac{\sigma^{y}}{E_{w}\alpha_{w}} \qquad [4.4]$$

This is called the upset zone.

Equation [4.1] together with the plastic properties during yielding can be integrated to give the axial stress. The solution can be found easily if the thermal expansion, Young's modulus and yield limit are constant. Figure 4.9 also contains a sketch of the behaviour for a material where Young's modulus, the yield

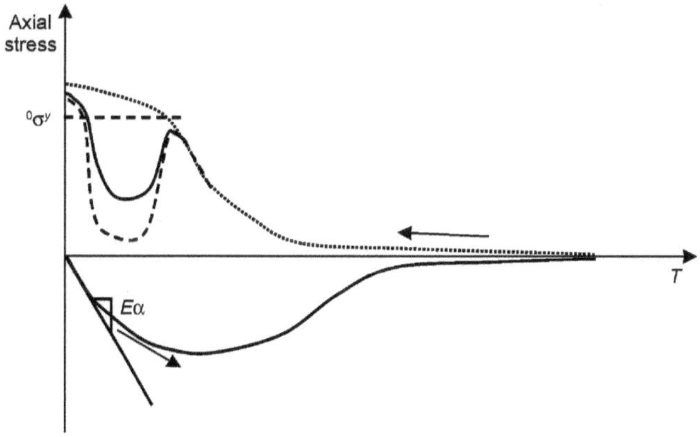

4.9 Simplified axial stress–temperature diagram for Satoh test. $^0\sigma^y$ is the virgin yield limit.

limit and hardening behaviour decrease with increasing temperature. The dashed curve indicates a possible stress evolution when a martensite formation occurs at lower temperature. The volume expansion will decrease the stress until it is completed. Thereafter the stress increases to a final residual value. The dotted curve shows the influence when the TRIP effect is neglected. A large tensile stress, up to yield limit of the material, is usually obtained but this may be changed by these phase transformations. They may even cause compressive longitudinal stresses in the weld centre [68]. The TRIP effect is explained in Section 9.7.2. Figure 4.10 shows the computed axial stress evolution [65] for a

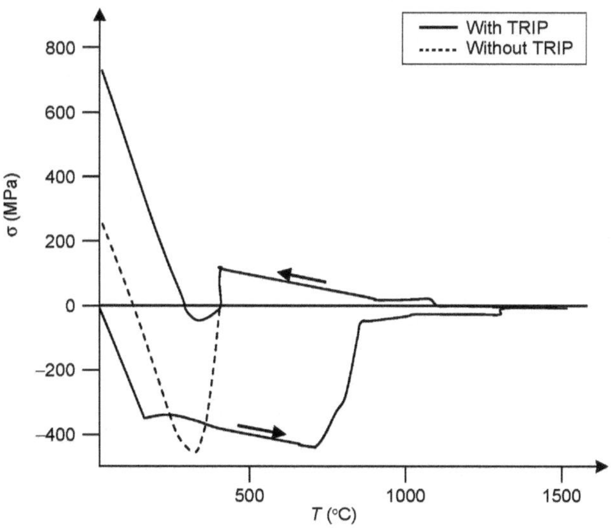

4.10 Computed stress–temperature diagram for Satoh test from Börjesson and Lindgren [65].

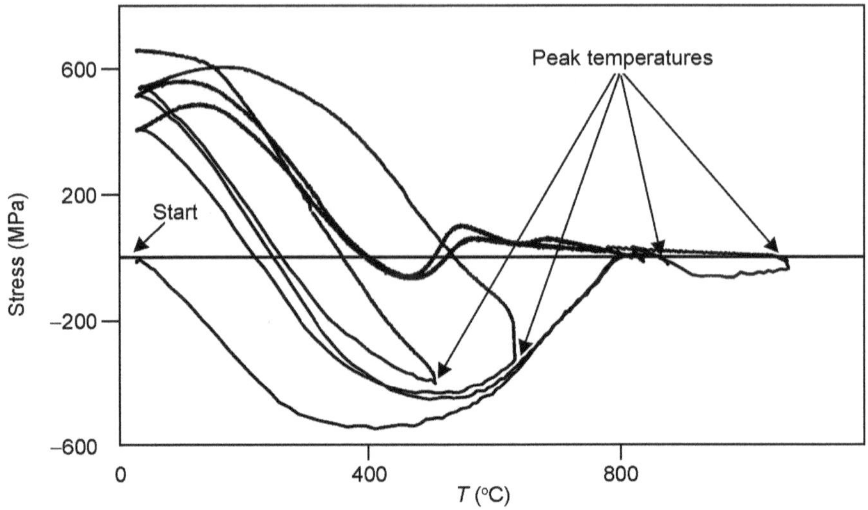

4.11 Measured stress–temperature for Satoh test with three thermal cycles, from Vincent *et al.* [69].

case with martensite formation where this is accounted for with and without the TRIP effect.

An example of a measured axial stress for a case where the temperature is cycled with peak temperatures of 1100, 900, 670 and 400 °C and cooling rate of 0.3 °C/s is shown in Fig. 4.11 (Vincent *et al.* [69]). The first thermal loading stage generates a completely bainitic transformation during cooling. The second gives a double ferritic-bainitic transformation.

The Satoh test is an extreme case of perfect rigid constraint, which is not possible in reality. The free thermal expansion gives deformations but no stresses. These two extremes show the problem in design of welding fixtures; large deformations or large stresses! It is very difficult to design a structure to reduce thermal deformations. The basic idea in the design of structures subjected to uneven temperatures is to allow for thermal expansion. Plastic deformations can then be avoided. The use of the Satoh test in calibration and validation is developed further in Section 8.5.1.

4.3 Welding of plate

The Satoh test does not illustrate the deformation behaviour during welding, only the stress evolution. It is the welding deformation that has often been the main concern in practical welding. The deformations may be so large that they are visible and cause misfit problems, and their correction can be costly [63]. It is easy to understand why welding deformation occurs but it is not easy to prevent it. The shrinkage of the weld both in the longitudinal and transverse directions is the cause. Case 1 in Fig. 4.12 shows the effect of the transverse

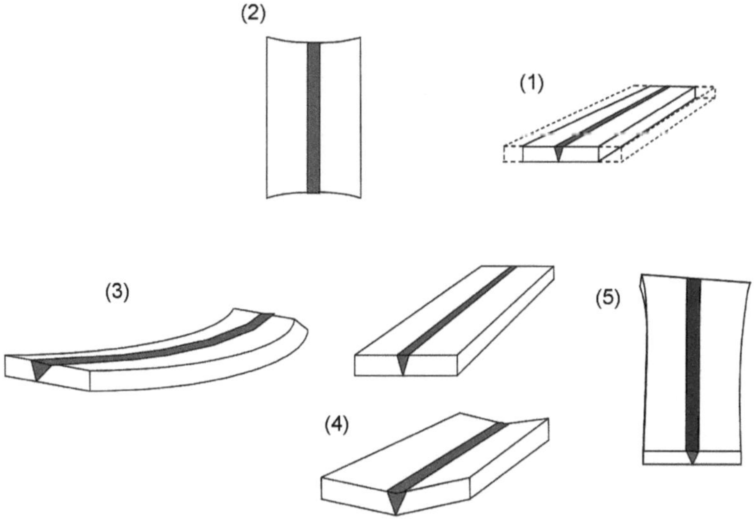

4.12 Deformation patterns.

shrinkage and case 2 is due to the longitudinal shrinkage. If the plate is thin relative to the weld size, then it may bow as the shrinkage is above the neutral axis of the plate as in case 3. A non-uniform transverse shrinkage over the thickness will give an angular distortion as in case 4. Clad plates may bend in two directions due to these shrinkages and produce a dished shape. Plates also dish inwards between stiffeners due to the weld shrinkages. Another deformation mode is called warping – case 5 in Fig. 4.12. The general view of plate deformation is usually based on the imagined case of a joint that is instantaneously filled with a molten metal (see [7, 59, 70]). This leads to a residual stress field of the type shown in Box 4.1 (see page 45). This is quite a typical case but phase changes may alter the residual stress distribution, even causing compressive longitudinal stresses in the weld centre [68]. The variations of the deformation behaviour can be even more pronounced as there are a number of contributing competitive factors (see Fig. 4.13). The weld pool is a soft region and does not directly affect the overall deformation. Although the laid weld is shrinking there is a certain volume that has been heated due to the weld. This volume contributes through its thermal expansion to the deformation. Then the joint ahead of the weld may have tack welds. These three regimes affect the overall deformations together with boundary conditions representing fixtures. The relation between the welding speed and the heat diffusivity is one major factor that can change the behaviour. Two different welding speeds are used for the simulations shown in Figs 4.14 and 4.15. The same heat input per unit length is used in the simulations. The peak temperature is higher and the whole temperature field is more focused near the arc in the case of a higher weld speed.

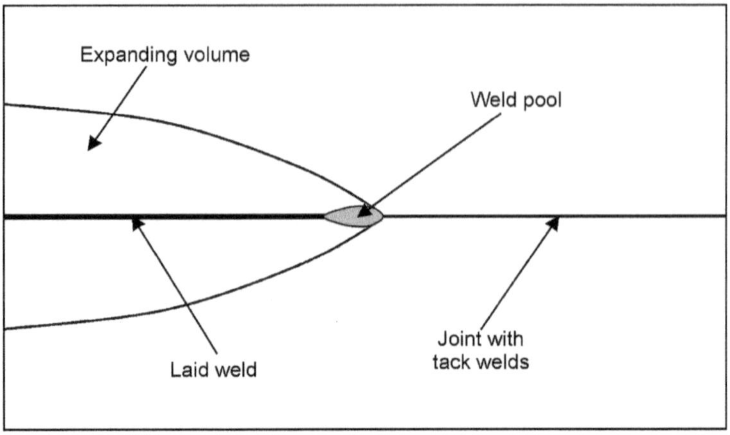

4.13 Domains contributing to plate deformation.

The Matlab code CWM_Lab has been used for the calculations below. There are small changes in the displacements field between when comparing the transverse displacements as well as (not shown) longitudinal displacement. The more localised temperature field gives a larger longitudinal expansion in the weld region. The region just outside the laid weld moves inward towards the weld. This region is smaller at higher speed. The plate moves more like a rigid body at the higher speed. Checking the numerical data for the transverse displacements at the lower edge of the joint shows the differences more clearly. The lower speed has a gap opening of 0.03 mm whereas the higher speed has a widening of 0.01 mm. The transverse shrinking at the start position of the weld is also somewhat smaller for the higher welding speed. It is 0.1 mm for the lower speed and 0.085 mm for the higher speed.

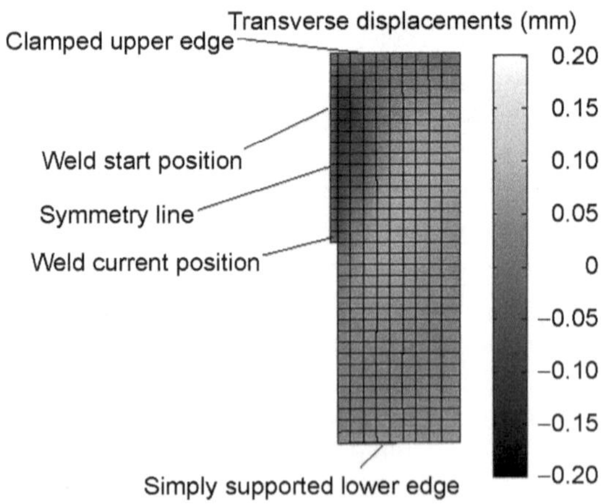

4.14 Transverse displacements of plate due to 'low' welding speed.

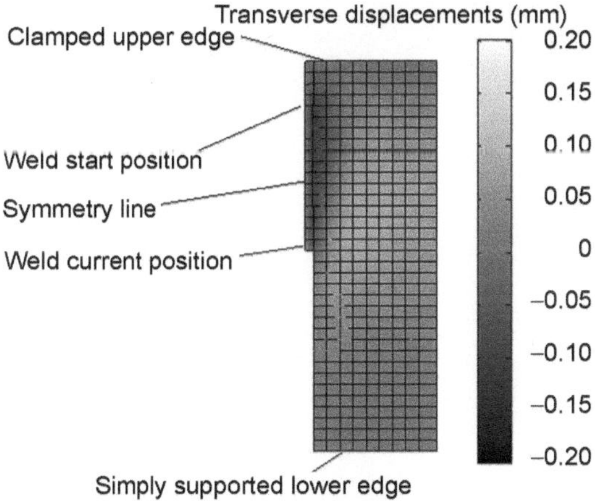

4.15 Transverse displacements of plate due to 'high' welding speed. The speed is two times higher than in Fig. 4.14.

The in-plane deformation in front of the welding arc may cause incomplete penetration if the gap between the plates to be welded closes too much. The opposite, a gap that opens too much, will cause burn-through. This was studied for a butt-welded plate with 0.01 m thickness and the final size of $1 \times 1\ m^2$. The measured and computed changes in gap width of about 0.12 m in front of the arc is shown in Fig. 4.16. The plate had four tack welds equally distributed along the plate length. The tack welds at the outer edges were laid first. Thereafter the two interior tack welds were made and their shrinkage created compressive forces on the two first tack welds. The gap decreased quickly at the start of the welding when the first tack weld was molten and its compressive forces were released. A gradual increase in the gap width built up due to the thermal expansion of the plate. This was counteracted by the last tack weld at the end of the welding.

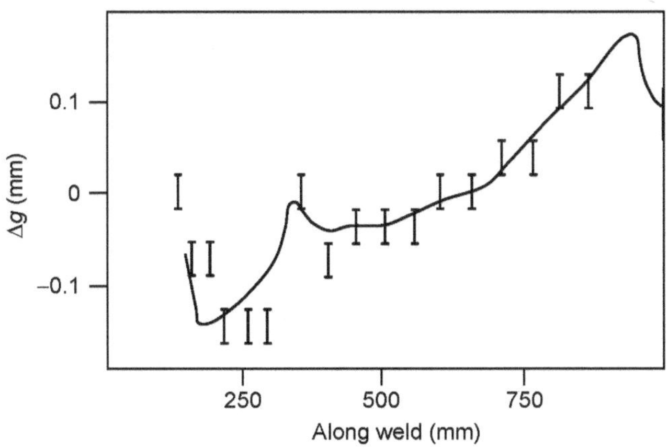

4.16 Change in gap width in front of arc [71].

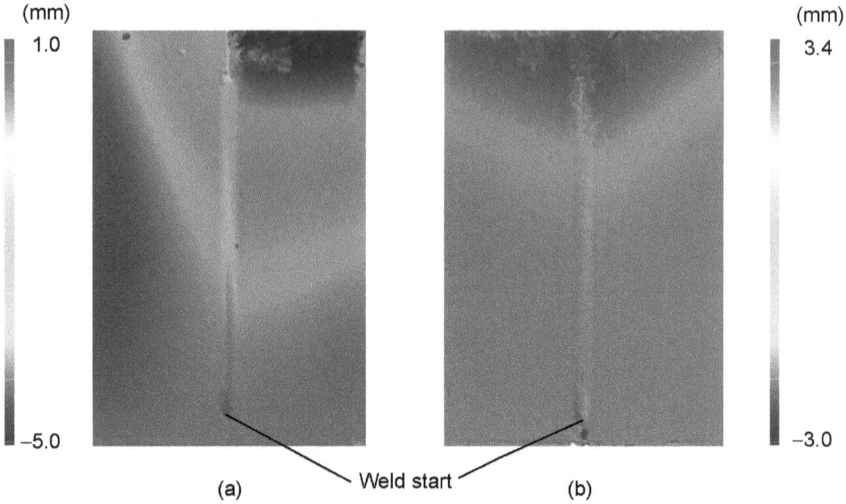

4.17 Deformation field for butt-welded plate. The plate is 50 × 100 × 1.7 mm^3.

There is a risk for failure at this weld due to the large tensile force it is subjected to. The figure illustrates how the order of tack welds affects the gap width.

The thinner a welded plate is in comparison with the size of the weld, the larger the deformations become. Thus the recommendation is to minimise the heat input per unit length of the weld. The use of deep penetration welds, such as electron beam (EB) welding or laser welding that have a reduced heat input, can still give problems for thin plate cases. Then the tensile residual stress along the weld may cause buckling. Nominally, identical cases may bow in different directions, which depend on small differences in the initial geometry. One example is shown in Fig. 4.17 where the measured out-of-plane deformations of two EB-welded plates are shown [72]. The weld condition and clamping is similar for both cases. They were clamped at the lower edge and a butt weld was performed. The start position is indicated on the plates and the welding was done 'upwards'. The deformation is not symmetric around the weld path for the left plate and a large mismatch between the individual halves is observed. This can be correlated to the initial mismatch, which was larger than for the right plate. Stone *et al.* [73] observed similar large variations for bowing of EB-welded plate. Dhingra and Murphy [74] obtained a good agreement between computed and measured deformation for thicker plates, 3 and 3.35 mm thick, but not for their thin plate, 1.5 mm thick. They attributed this to buckling. Issues concerning robustness and stability are discussed in Chapter 11.

The used of balanced placing of weld passes, appropriate use of restraints including tack welds, minimising heat input, etc., are attempts to reduce these deformations. On the other hand, more restraints increase the welding stresses and other problems may appear. The optimal solution would be to preform the welded component so that the required shape after welding is obtained.

Box 4.1 Butt-welded plate

The plate is clamped at the upper edge in Figs 4.18 and 4.19. There is a slit from weld start position which is filled during welding, leaving the remaining part of the gap open at finish of welding.

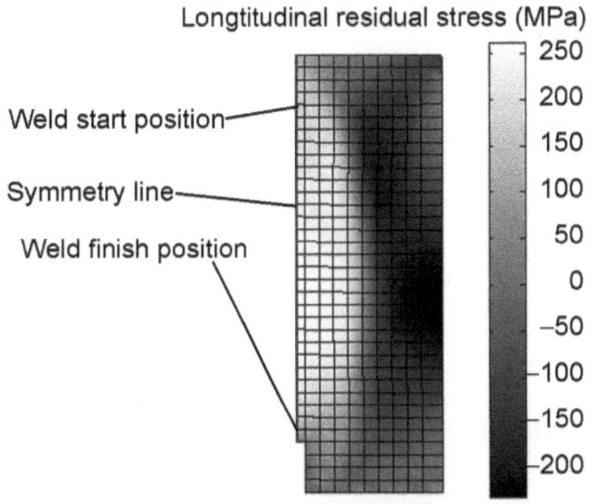

4.18 Longitudinal stress field for bead on plate case.

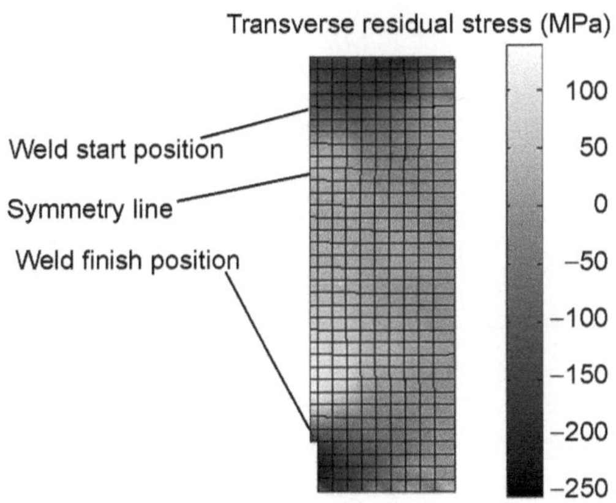

4.19 Transverse stress field for bead on plate case.

4.4 Welding of pipe

Circumferential welding of pipes is similar to welding of plates. The longitudinal residual stress gives a radial shrinkage in the weld. The curvature of the pipe gives a more controlled initial geometry with less risk for instability during welding. The shrinkage in the weld reduces the pipe diameter. The pipe in Fig. 4.20 was studied by Karlsson and Josefson [75], Josefson and Karlsson [76] and Murthy *et al.* [77]. The measured and computed reduction in diameter

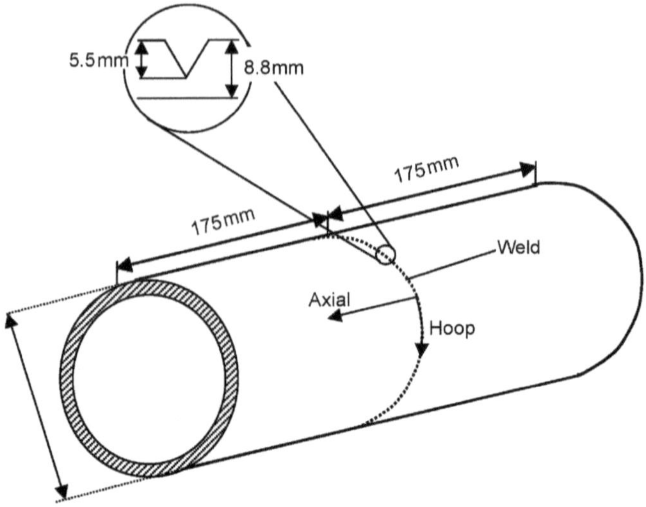

4.20 Butt-welded pipe [75–77].

is shown in Fig. 4.21. The measurements show an asymmetric deformation both in the circumferential direction as well as transverse to the weld line. The latter may be because the clamping of the pipe was applied only at one end. The modelling of circumferential welding of pipes using axisymmetric or fully 3D shells or solid element models is discussed in Section 7.6.

Plates usually form a load-carrying structure together with stiffeners or other restraints. Welded pipes may need to be able to carry a load as without additional stiffeners. The welding residual stresses and especially the deformations will then reduce the buckling strength [78].

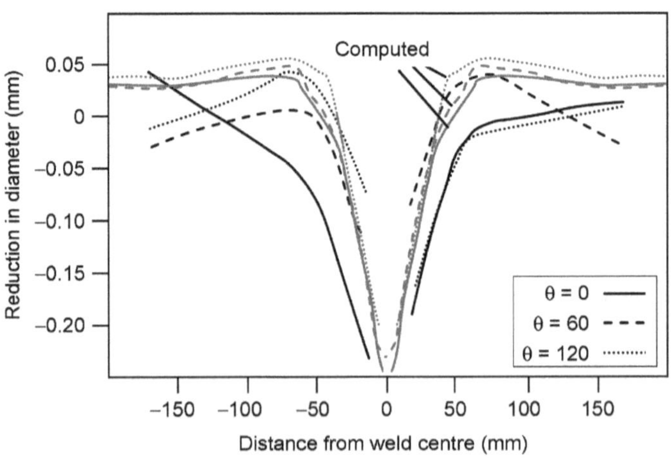

4.21 Change in diameter for butt-welded pipe along lines with different angles with respect to start position [78].

5
Nonlinear heat flow

The fundamental relations in heat conduction of a rigid heat conductor are summarised below and the corresponding finite element relations are given. The solution of the nonlinear heat conduction in CWM problems is fairly straightforward compared with the solution of nonlinear deformations. The sizes of the models are smaller as the temperature is a scalar field whereas the displacement is a vector field. Furthermore, the nonlinearities are less troublesome. The only issue that may require some caution is the latent heat due to phase changes, especially the heat of fusion. Different numerical techniques to handle this nonlinearity are described.

5.1 Basic equations of nonlinear heat conduction

The energy balance between change in stored energy and heat flux leads to [24, 30, 79]

$$\rho \dot{H} = \dot{Q} - \nabla \boldsymbol{q} \qquad [5.1]$$

where $\dot{Q}$ [W/m^3] is the power per unit volume, ρ is the density, $\dot{H}$ [J/kg] is the volumetric enthalpy or heat content and $\boldsymbol{q}$ [W/m^2] is the heat flux vector. Fourier's law for isotropic heat conduction gives

$$\boldsymbol{q} = -\boldsymbol{\lambda} \nabla T \qquad [5.2]$$

where $\boldsymbol{\lambda}$ [W/m °C] or [W/m K] is a diagonal matrix with different conductivities in each direction in case of anisotropic heat conduction. In the following, the focus is on isotropic heat conduction and use

$$\boldsymbol{q} = -\lambda \nabla T \qquad [5.3]$$

where λ now is the heat conductivity number of the material.

Fourier's law is an approximation implying an infinite speed of heat conduction [80, 81]. However, the heat conduction is a slow process even with this model, as the amount of heat conducted far away from a heat source in a short time is very small. It is customary to use the notation of thermal diffusivity

$$a = \frac{\lambda}{\rho c} \qquad [5.4]$$

as a measure of how fast heat is conducted in a solid.

The enthalpy is related to the temperature by

$$H(T) = \int_{T_{\text{ref}}}^{T} c(\tau)\mathrm{d}\tau \qquad [5.5]$$

where c is the heat capacity [J/kg °C]. This equation may be used to define heat capacity as

$$c = \frac{\mathrm{d}H}{\mathrm{d}T} \qquad [5.6]$$

The relations above can be combined to produce the classical heat conduction equation

$$\rho c\dot{T} = \dot{Q} + \nabla(\lambda \nabla T) \qquad [5.7]$$

Some finite element formulations that are introduced to reduce the numerical problems in presence of latent heats are based on the enthalpy version [24]

$$\rho \dot{H} = \dot{Q} + \nabla(\lambda \nabla T) \qquad [5.8]$$

or an enthalpy gradient version [82]

$$\rho \dot{H} = \dot{Q} + \nabla\left(\frac{\lambda}{c} \nabla H\right) \qquad [5.9]$$

Equation [5.7] can be rewritten using Eq. [5.6]

$$\rho \frac{\mathrm{d}H}{\mathrm{d}T}\dot{T} = \dot{Q} + \nabla(\lambda \nabla T) \qquad [5.10]$$

known as the enhanced or apparent heat capacity equation.

The heat conduction equation, together with initial and boundary conditions, defines the problem to be solved. Simple boundary conditions are prescribed temperature or prescribed heat flux, q_{n} is defined as positive when directed in the outward normal direction. It is zero in case of an isolated, adiabatic, boundary. Convective and radiation heat losses are more complex boundary conditions for the outward flux. Then the flux depends on the temperature of the body and the surrounding and is written as

$$q_{\text{n}} = -\lambda \nabla T \cdot \boldsymbol{n} = -\lambda \frac{\partial T}{\partial \boldsymbol{n}} = \alpha(T - T_{\infty}) + \epsilon\sigma(T^4 - T_{\infty}^4) \qquad [5.11]$$

where the first term is convective heat loss and α is the heat transfer coefficient. The second term is the heat loss due to radiation and σ is Stefan–Boltzmann's constant and ϵ is the emissivity factor. The latter is 1 for a perfect black body. The latter term is a nonlinear boundary condition. The equation can be rewritten as

$$q_n = \left[\alpha + \epsilon\sigma(T^2 + T_\infty^2)(T + T_\infty)\right](T - T_\infty) = \alpha_{\text{eff}}(T - T_\infty) \qquad [5.12]$$

See Section 9.11 for further information about these surface properties.

5.2 Finite element formulation of nonlinear heat conduction

Front tracking techniques can be used for detailed analysis of problems with phase changes. The extended finite element method (XFEM) has also been used for these kinds of problem. However, there are simpler approaches that work satisfactorily for welding problems. Thus the following focuses on a fixed mesh method based on the enhanced heat capacity version of the heat conduction equation, Eq. [5.10].

The standard finite element semidiscretisation procedures for the heat conduction equation [24, 32, 83] lead to an expression for nodal energy equilibrium

$$\boldsymbol{C}\dot{\boldsymbol{T}} = \dot{\boldsymbol{Q}}_{\text{ext}} - \dot{\boldsymbol{Q}}_{\text{int}} \qquad [5.13]$$

where $\boldsymbol{C}$ is the heat capacity matrix, $\dot{Q}_{\text{int}}$ is the internal flux vector and $\dot{Q}_{\text{ext}}$ is the thermal load vector. The latter may include plastic dissipated heat and the effect of elastic strains from the mechanical analysis in a strongly coupled analysis solved by, for example, a staggered approach.

The temperature field within each element is interpolated as

$$^{n+\theta}_{\ \ i}T(\boldsymbol{x}, t) = \boldsymbol{N}_{\text{th}}(\boldsymbol{x})^{n+\theta}_{\ \ i}\boldsymbol{T}^{\text{e}}(t) \qquad [5.14]$$

where $\boldsymbol{N}_{\text{th}}(\boldsymbol{x})$ is a matrix with the interpolation (shape) functions and $^{n+\theta}_{\ \ i}\boldsymbol{T}^{\text{e}}(t)$ is the current estimate of the element temperature. The variable θ is explained later. The left superscript n is the number of the time increment and the left subscript is an iteration counter. They will be used later. The coordinate $\mathbf{x}$ is assumed to be constant during the increment (rigid conductor). In the staggered approach (Chapter 3), it will be the coordinate at the start of the increment, $^n\boldsymbol{x}$.

The gradient of the temperature field is

$$^{n+\theta}_{\ \ i}\boldsymbol{g} = {}^{n+\theta}_{\ \ i}\nabla\boldsymbol{T} = \nabla\boldsymbol{N}_{\text{th}}\,{}^{n+\theta}_{\ \ i}\boldsymbol{T}^{\text{e}}(t) = \boldsymbol{B}_{\text{th}}\,{}^{n+\theta}_{\ \ i}\boldsymbol{T}^{\text{e}} \qquad [5.15]$$

where the matrix $\boldsymbol{B}_{\text{th}}$ has the first derivatives of the interpolation functions.

The element matrices $\boldsymbol{N}_{\text{th}}$, $\boldsymbol{B}_{\text{th}}$ and λ and the density ρ and heat capacity c are used to compute element contributions to Eq. [5.13]. We have

$$^{n+\theta}_{\ \ i}\boldsymbol{c} = \int_{v^{\text{e}}} \rho c\left({}^{n+\theta}_{\ \ i}T\right)\boldsymbol{N}_{\text{th}}^T\boldsymbol{N}_{\text{th}}\,\text{d}v \xrightarrow{\text{assembling}} {}^{n+\theta}_{\ \ i}\boldsymbol{C} \qquad [5.16]$$

and the nodal heat flux and Fourier's equation give

$$
{}_{i}^{n+\theta}\dot{\boldsymbol{q}}_{\text{int}} = \int_{v^{\text{e}}} \boldsymbol{B}_{\text{th}}^{T} \lambda\left({}_{i}^{n+\theta}T\right) \boldsymbol{B}_{\text{th}} \mathrm{d}v_{i}^{n+\theta} \boldsymbol{T}^{(\text{e})}
$$

$$
= {}_{i}^{n+\theta}\boldsymbol{k}_{\text{th}}\, {}_{i}^{n+\theta}\boldsymbol{T}^{(\text{e})} \xrightarrow{\text{assembling}} {}_{i}^{n+\theta}\dot{\boldsymbol{Q}}_{\text{int}} = {}_{i}^{n+\theta}\boldsymbol{K}_{\text{th}}\, {}_{i}^{n+\theta}\boldsymbol{T} \qquad [5.17]
$$

where ${}_{i}^{n+\theta}\boldsymbol{K}_{\text{th}}$ is the conductivity matrix and v^{e} is the volume of the element, i.e. at start of increment in the staggered approach. The details of this are given in the literature for finite element methods, e.g. [24]. It is implemented in CWM_Lab in LEL_Ther_El_Loop.

The load vector ${}_{i}^{n+\theta}\dot{\boldsymbol{Q}}_{\text{ext}}$ can have element contributions for volumetric heating as well as from nodal or surface heat flux. Boundary conditions that are dependent on the temperature of the body, such as convection and radiation, will also contribute to the heat conductivity matrix [83].

The finite element semidiscretisation leads thus to element contributions assembled to the nonlinear coupled system of equations (Eq. [5.13]). The temporal discretisation of this system is often based on the generalised midpoint method. It aims at fulfilling these energy equilibrium equations at time

$$
{}_{i}^{n+\theta}t = {}^{n}t + \theta\Delta t \qquad [5.18]
$$

giving

$$
{}_{i}^{n+\theta}\boldsymbol{C}_{i}^{n+\theta}\dot{\boldsymbol{T}} = {}_{i}^{n+\theta}\dot{\boldsymbol{Q}}_{\text{ext}} - {}_{i}^{n+\theta}\dot{\boldsymbol{Q}}_{\text{int}} \qquad [5.19]
$$

where n is the increment/time step number, ${}^{n}t$ is the time at the start of the time step, Δt is the length of the current time step and the parameter $\theta \in [0, 1]$ determines where in the increment we want to fulfil the energy equilibrium. The left subscript i is the iteration counter as explained soon.

The method is unconditionally stable if $\theta \geq 0.5$. The use of 0.5 gives a second order accurate time stepping but the solution can show oscillations. Further details about stability of the time stepping can be found in Lewis *et al.* [83], Belytscho [84] and Hughes [85]. The author prefers to use as stable approach as possible, i.e. $\theta = 1$. Some researchers have used explicit time stepping schemes that can be efficiently combined with lumped heat capacity matrices. However, they did not continue using this approach owing to the conditional stability criterion preventing the use of long time steps in the cooling phase (see Lindgren [86] for references). The explicit approach was used successfully by Bergman and Oldenburg [87] for sheet metal forming with controlled temperature.

The nonlinearities in Eq. [5.19] require an iterative approach for solving the temperature at each time step. A predictor is used for the first iteration and thereafter an iterative corrector phase is needed. The predictor can be as simple as assuming no change in temperature during the time step. The corrector phase aims to fulfil Eq. [5.13] at time ${}^{n+\theta}t$. We rewrite the equation for this purpose to

$$
{}_{i+1}\boldsymbol{R}_{\text{th}} = \left[{}_{i+1}^{n+\theta}\dot{\boldsymbol{Q}}_{\text{ext}} - {}_{i+1}^{n+\theta}\boldsymbol{C}_{i+1}^{n+\theta}\dot{\boldsymbol{T}} - {}_{i+1}^{n+\theta}\dot{\boldsymbol{Q}}_{\text{int}}\right]\Delta t \qquad [5.20]
$$

where the left subscript is the iteration counter.

The midpoint temperature is

$$ {}_{i}^{n+\theta}\boldsymbol{T} = {}^{n}\boldsymbol{T} + \theta_{i}\Delta\boldsymbol{T} \qquad [5.21] $$

and we introduce iterative corrections to the increment in temperature as

$$ {}_{i+1}\Delta\boldsymbol{T} = {}_{i}\Delta\boldsymbol{T} + {}_{i}\delta\boldsymbol{T} \qquad [5.22] $$

The midpoint method uses

$$ {}_{i+1}^{n+\theta}\dot{\boldsymbol{T}} = \frac{{}_{i+1}\Delta\boldsymbol{T}}{\Delta t} = {}_{i}^{n+\theta}\dot{\boldsymbol{T}} + \frac{{}_{i}\delta\boldsymbol{T}}{\Delta t} \qquad [5.23] $$

The Newton–Raphson iterative approach is based on a Taylor expansion of Eq. [5.21] where we want to find the iterative correction that may give ${}_{i+1}^{n+\theta}\boldsymbol{R} = \boldsymbol{0}$ in the next iteration.

$$ {}_{i+1}^{n+\theta}\boldsymbol{R}_{\text{th}} \approx {}_{i}^{n+\theta}\boldsymbol{R}_{\text{th}} + \frac{\partial_{i}^{n+\theta}\boldsymbol{R}_{\text{th}}}{\partial\boldsymbol{T}}\,{}_{i}\delta\boldsymbol{T} = \boldsymbol{0} \qquad [5.24] $$

This can be written as

$$ -\frac{\partial_{i}^{n+\theta}\boldsymbol{R}_{\text{th}}}{\partial\boldsymbol{T}}\,{}_{i}\delta\boldsymbol{T} = {}_{i}^{n+\theta}\boldsymbol{R}_{\text{th}} \qquad [5.25] $$

or introducing the tangent matrix

$$ {}_{i}^{n+\theta}\boldsymbol{C}_{\text{t}}\,{}_{i}\delta\boldsymbol{T} = {}_{i}^{n+\theta}\boldsymbol{R}_{\text{th}} \qquad [5.26] $$

The tangent matrix is needed to retain second order convergence of the Newton–Raphson method in this iterative process. It becomes

$$ {}_{i}^{n+\theta}\boldsymbol{C}_{\text{t}} = -\frac{\partial_{i}^{n+\theta}\boldsymbol{R}_{\text{th}}}{\partial\boldsymbol{T}} = \left[\frac{\partial\left({}_{i}^{n+\theta}\boldsymbol{C}_{i}^{n+\theta}\dot{\boldsymbol{T}}\right)}{\partial\boldsymbol{T}} + \frac{\partial_{i}^{n+\theta}\dot{\boldsymbol{Q}}_{\text{int}}}{\partial\boldsymbol{T}} - \frac{\partial_{i}^{n+\theta}\dot{\boldsymbol{Q}}_{\text{ext}}}{\partial\boldsymbol{T}}\right]\Delta t \qquad [5.27] $$

and it is usually approximated to

$$ {}_{i}^{n+\theta}\boldsymbol{C}_{\text{t}} \approx \left[{}_{i}^{n+\theta}\boldsymbol{C}\frac{1}{\Delta t} + {}_{i}^{n+\theta}\boldsymbol{K}_{\text{th}}\theta\right]\Delta t = {}_{i}^{n+\theta}\boldsymbol{C} + \theta\Delta t_{i}^{n+\theta}\boldsymbol{K}_{\text{th}} \qquad [5.28] $$

A modified Newton–Raphson approach may be more efficient. Then the tangent matrix is not updated every iteration. Typically it is updated during the first few iterations and thereafter only if divergence occurs. The incremental, iterative approach for the nonlinear heat conduction problem is summarised in Fig. 5.1.

The thermal analysis in welding simulations is usually straightforward. The size of the problem is smaller than for the mechanical part, as there is only one unknown per node. The material properties are also easier to find and the numerical problem in the nonlinear solution procedure is smaller. However, the latent heats due to phase changes can make the heat conduction equation stiff. The enthalpy method [71, 88, 89] can be used where the enthalpy, heat content,

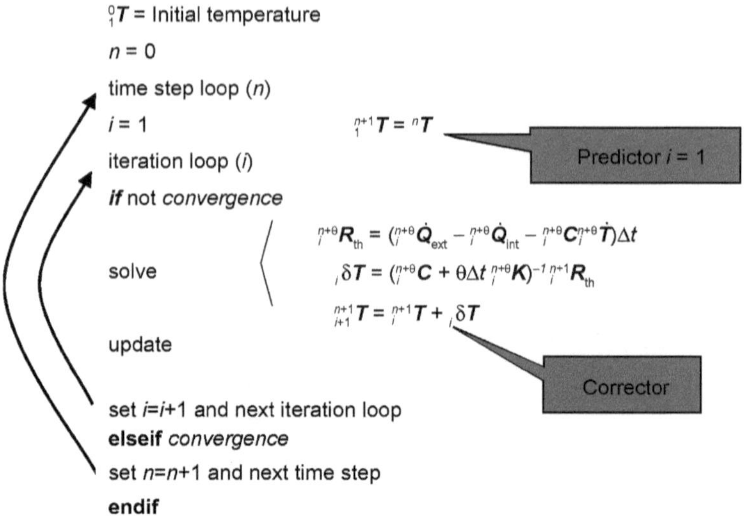

5.1 Incremental, iterative procedure for thermal analysis in function LEL_Thermod in CWM_Lab.

is used as the primary unknown in the finite element solution. It is defined as

$$H(T) = \int_{T_{\text{ref}}}^{T} \rho c(\tau)\mathrm{d}\tau \qquad [5.29]$$

This is a monotonous increasing function with respect to the temperature and will reduce the nonlinearity when latent heats are consumed/released. Another numerical approach is the fictitious heat source method [90] used by Murthy *et al.* [77]. A third approach, used by, for example, Lindgren *et al.* [3], is based on using an enhanced, regularised, heat capacity [83]. It is sometimes also called an enthalpy method. The heat capacity in Eq. [5.16] is then taken as

$$^{n+\theta}c = c_{\text{eff}} = \frac{{}^{n+1}_{i}H - {}^{n}H}{\Delta t} \qquad \text{if } \Delta t \neq 0 \qquad [5.30]$$

or in the case of severe convergence problems

$$^{n+\theta}c = c_{\text{eff}} = \frac{{}^{n}H - {}^{n-1}H}{{}^{n}t - {}^{n-1}t} \qquad [5.31]$$

The method works well for alloys where the melting occurs over a temperature range. The most straightforward way to reduce the nonlinearity is to extend the temperature range over which the latent heat is released/consumed in the model.

Pham [82] compared temperature gradient and enthalpy gradient formulations combined with lumped or distributed capacitance matrices. The physical interpretation of lumped matrices is that all heat storage occurs at the nodes. They are diagonal matrices. Pham also compared different methods to compute apparent heat capacity. He found that time averaging methods as in Eqs [5.30] or [5.31] should be implemented with a lumped heat capacity formulation. Other relative merits of lumped versus distributed heat capacities is that the former has a better stability range and resistance to oscillation [91]. The latter reduces the overshoot in the numerical solution that may occur in front of the arc where the steep gradient may give a temperature that is lower than the initial temperature.

The formulation given above is demonstrated on a one-dimensional Stefan problem in Box 5.1. The numerical solution is compared with an analytical solution in Carslaw and Jaeger [30] for a semi-infinite long bar.

Box 5.1 Heat conduction problem with latent heat

A rod with zero initial temperature has a latent heat $L = 70.26$, $T_{\text{liquidus}} = -0.15$ and $T_{\text{solidus}} = -10.15$. The left end is initially given the temperature $T_{\text{left}} = -45$. All other properties are given unit value. The solution based on the apparent heat capacity approach in Eq. [5.31] is used and the solution logic in Fig. 5.1 is used. The results are compared with data in Lewis *et al.* [83]. The solution in Fig. 5.2 is based on 40 linear, two-node elements of equal length and 100 equal long time steps. A lumped heat capacity matrix was used. The current case gave the same results at $x = 1$ for a consistent heat capacity matrix.

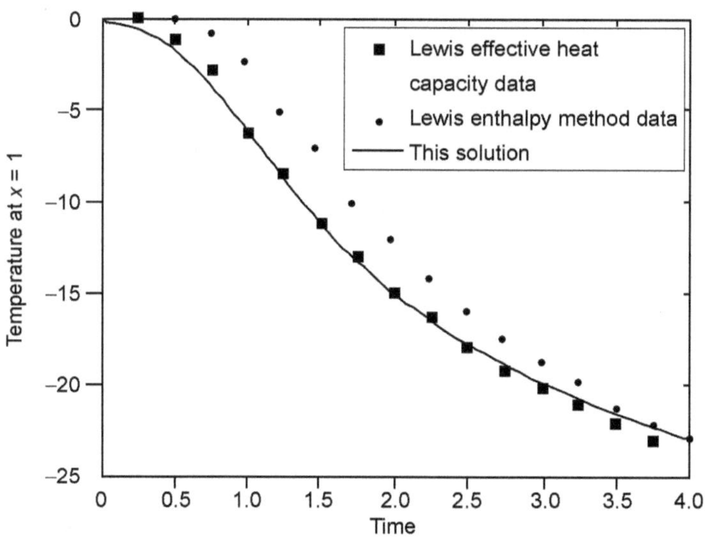

5.2 Temperature versus time for $x = 1.0$.

6
Nonlinear deformation

Welding deformations may quite often be visible to the naked eye, implying that large deformation and rotation effects are important. The strains are usually not so large except in the weld pool but this region is simplified considerably in the classical CWM approach (see Section 3.3). However, the additional cost of accounting for large strains is small. Accounting for large deformations makes the mechanical analysis much more complex than the thermal analysis and there are several options for how to treat them. It is possible to use different mesh descriptions, different kinetic and kinematic variables, and constitutive models (see [32, 34, 92–95]).

The purpose of this chapter is to give an overview of the solution algorithm for nonlinear deformations. The interested reader can use the chapter as an introduction and then study the details of the derivations in books devoted to nonlinear mechanics and finite element formulation, e.g. [34]. Thereafter, it would be advisable to return to the CWM software associated with this book (see pages xiii–xiv) in order to master one approach for finite element implementation of the hyperelastoplastic model and in the framework of large deformations and thermomechanics. The latter use a staggered approach (Section 3.3).

The focus is on the updated Lagrangian approach as it is the most common method together with the concepts of Cauchy true stress and true logarithmic strain. Two different approaches for the constitutive modelling will be outlined; hypoelastic and hyperelastic formulations. The first is more common but the second is more correct. They give the same results for metal plasticity. Both methods can be formulated so that the stress updating can be done with the same logic. The chapter concludes with a description of a stress-updating algorithm for deviatoric plasticity, the radial return method.

A quasistatic mechanical analysis is sufficient as the inertia effects in fusion welding processes are negligible, as stated in Section 3.3. Thus time stepping methods such as Newmark's algorithm and mass matrices will not be discussed. The interested reader is referred to Belytschko *et al.* [34] and Belytschko and Hughes [96].

6.1 Basic choices in formulation of nonlinear deformation

The basic equations of nonlinear deformations [28, 32, 34, 92–95, 97, 98] are the equilibrium equations (sometimes called the kinetic relations), constitutive stress–strain relations and geometric compatibility (or kinematic), relations as shown in Fig. 6.1. The notations in the figure are introduced later.

There are several options for these basic relations:

- Reference configuration and corresponding stress and strain definitions.
- Reference frame for the mesh.
- Approach for the formulation of the constitutive model.

The reference configuration is the geometry to which all variables are associated. There are two basic options: original or current geometry; the latter is discussed here. Bonet and Wood [94] described the two variants side by side. The Cauchy true stress and the logarithmic true strain definitions of stress and strain will be used in this context. They are illustrated in the uniaxial case in Fig. 6.2. The displacement u gives a strain in the rod and several definitions are possible. The logarithmic strain definition is described below. The idea is that the strain increment at each instance during the loading is the increment in strain divided by the current length

$$d\epsilon = \frac{du}{L_0 + u} = \frac{\dot{u}dt}{L} = \frac{\dot{L}}{L} \cdot dt = d \cdot dt \qquad [6.1]$$

The term d is also introduced above. It is the one-dimensional variant of the spatial velocity gradient used later. The total strain is the sum, integral, of all

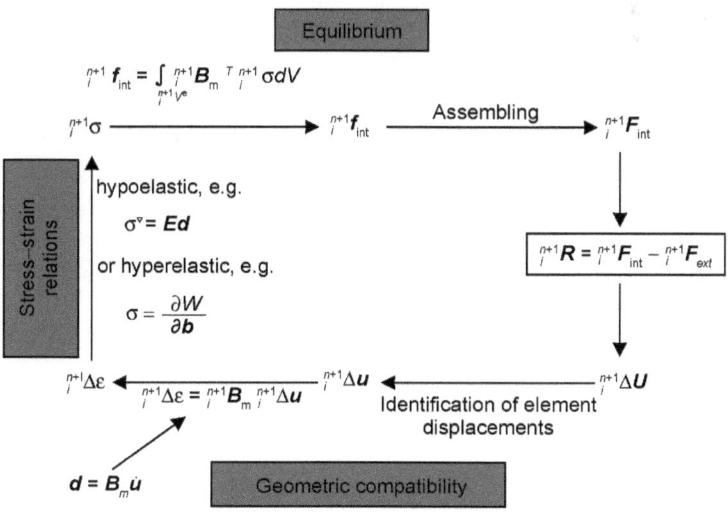

6.1 Tönti diagram for basic relations in nonlinear deformation.

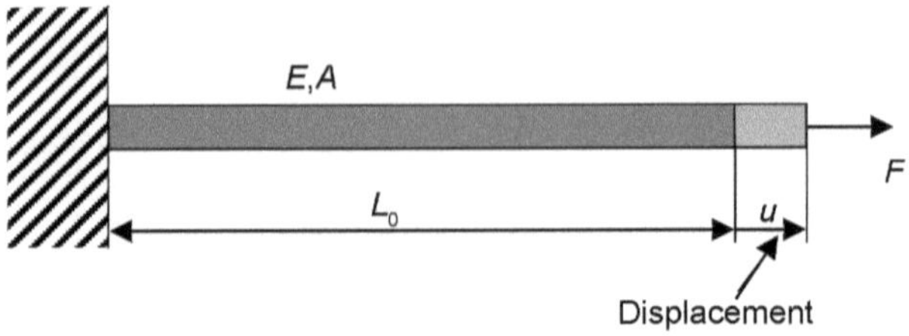

6.2 Uniaxial deformation.

these infinitesimal increments

$$\epsilon = \int_0^u \frac{d\kappa}{L_0 + \kappa} = \int_0^u \frac{1}{L_0} \frac{d\kappa}{1 + \kappa/L_0} = \left[\ln\left(1 + \frac{\kappa}{L_0}\right)\right]_0^u$$

$$= \ln\left(1 + \frac{u}{L_0}\right) - \ln(1) = \ln(\lambda) \qquad [6.2]$$

where the stretch

$$\lambda = 1 + \frac{u}{L_0} = \frac{L_0 + u}{L_0} = \frac{L}{L_0} \qquad [6.3]$$

has been introduced.

This is the type of strain definition used in the following discussion and the reason for the name 'logarithmic strain' is obvious. Sometimes it is called natural strain definition. There is a relation between the rate of the stretch and the velocity strain

$$d = \frac{\dot{L}}{L} = \frac{\dot{L}}{L_0} \frac{L_0}{L} = \dot{\lambda} \frac{L_0}{L} = \dot{\lambda} \frac{1}{\lambda} \qquad [6.4]$$

The same result is obtained by taking the time derivative of Eq. [6.2].

There is a conjugate stress measure to each strain definition. The product of these measures should be work. The choice of reference configuration must be done when the conjugate stress measure is derived. Current geometry is used here. The virtual work done by the external force, F, when the right edge of the rod is given a virtual displacement, δu will equal the internal virtual work done by the stresses. Integrate over the current volume and introduce a as current cross-sectional area. The internal virtual work is

$$\delta V_{\text{int}} = \int_0^L \sigma \delta \epsilon a dx = \int_0^L \sigma \frac{\delta \lambda}{\lambda} a dx = \sigma \frac{\delta \lambda}{\lambda} aL \qquad [6.5]$$

and the external virtual work is

$$\delta V_{\text{ext}} = F \delta u \qquad [6.6]$$

Setting these equal gives

$$\sigma \frac{\delta\lambda}{\lambda} aL = F\delta u \rightarrow \sigma\delta\lambda = \frac{F}{a}\lambda\frac{\delta u}{L} \rightarrow \sigma\delta\lambda = \frac{F}{a}\frac{\delta u}{L_0} \rightarrow \sigma\delta\lambda = \frac{F}{a}\delta\lambda \quad [6.7]$$

Thus the Cauchy true stress definition is obtained as

$$\sigma = \frac{F}{a} \quad [6.8]$$

The reference frame used here for the mesh is the Lagrangian approach [34], where each node in the mesh is connected to a material point. The alternative is the Eulerian frame where the mesh is fixed and the material flows through the mesh. There exists an arbitrarily Lagrangian Euler (ALE) mesh description which is a combination of both variants. This is connected with remeshing discussed in Section 7.3. The discussions are based on a variant of the Lagrangian mesh called updated Lagrangian (UL) formulation. The mesh is updated at each time step. Thus the ${}^{n}t$ configuration replaces the original configuration as the reference state (Fig. 6.3). It shows the motion; some of the notations in the figure are discussed later. The increment from this state to state ${}^{n+1}t$ is to be obtained as described in the next section.

The most important concept to understand is the deformation gradient $\boldsymbol{F}$. It is a two-point tensor mapping neighbourhoods from original to current configuration

$$d\boldsymbol{x} = \boldsymbol{F}\mathrm{d}\boldsymbol{X} \quad [6.9]$$

Thus it includes rigid body rotation and deformation of these line segments (see Fig. 6.3). Thereby it contains all the information needed to determine different strain measures and also to relate quantities in original and current configuration. One example is the Euler strain definition, $\boldsymbol{e}_{\mathrm{E}}$, obtained from

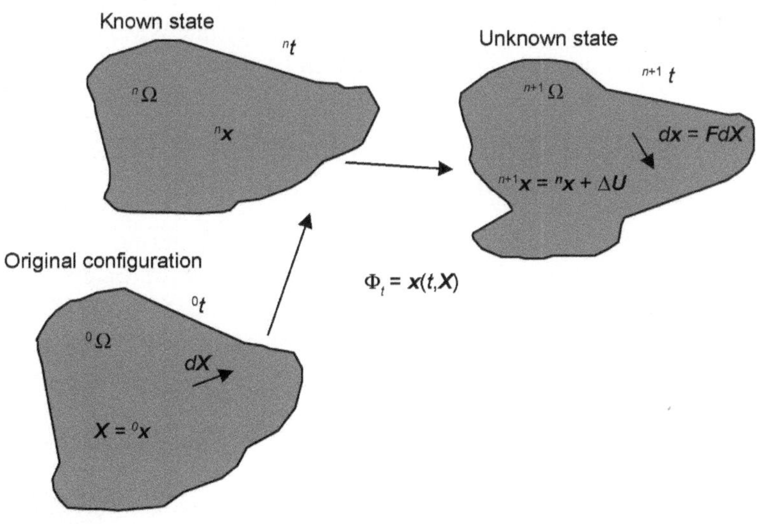

6.3 Basis for updated Lagrangian mesh description.

$$\frac{ds^2 - dS^2}{2} = \frac{d\boldsymbol{x}^T d\boldsymbol{x} - d\boldsymbol{X}^T d\boldsymbol{X}}{2} = d\boldsymbol{x}^T \frac{\boldsymbol{I} - \boldsymbol{b}^{-1}}{2} d\boldsymbol{x} = d\boldsymbol{x}^T \boldsymbol{e}_E d\boldsymbol{x} \qquad [6.10]$$

Thus the left Cauchy–Green or Finger tensor, $\boldsymbol{b}$, is also a measure of the total deformation in the current configuration. It is

$$\boldsymbol{b} = \boldsymbol{F}\boldsymbol{F}^T \qquad [6.11]$$

The deformation gradient plays an important role in the mapping of different quantities between different configurations [34, 94, 95]. It is identical to the stretch in the uniaxial case above, as we have

$$dx = \lambda dX \qquad [6.12]$$

It is possible to use a so-called flow formulation where the material is treated as a non-Newtonian fluid. It is often combined with an Eulerian reference frame. The latter is explained below. The solid formulation, i.e. the use of the constitutive models for solids, is focused on here. The options are:

- hypoelastic constitutive model;
- hyperelastic constitutive model.

Both variants are discussed below and formulated so that the same logic for stress-updating can be used.

Section 6.2 gives a formulation of the overall, global, solution strategy for the nonlinear deformation problem. It is quite general, although it is expressed in the current configuration, as integrals are used over the current volume as well as Cauchy stress and logarithmic strain definitions. It can be translated to other approaches by changing the interpretation of these variables. Thereafter Section 6.3 describes the hypoelastic and hyperelastic formulation for plasticity problems and Section 6.4 describes the radial return algorithm. The motivation for the latter is that it may sometimes be necessary in CWM to implement a user material model and this approach can then be used.

6.2 Finite element formulation of nonlinear deformation

The principle of virtual power is often the starting point for formulation of the finite element method. This corresponds to the use of Galerkin methods, a special case of weighted residual methods. This leads to (e.g. [32, 34, 93, 94]) the coupled system of equations for the motion of the nodes in the finite element mesh

$$\boldsymbol{M}\ddot{\boldsymbol{U}} \approx \boldsymbol{0} = \boldsymbol{F}_{ext} - \boldsymbol{F}_{int} \qquad [6.13]$$

where $\boldsymbol{M}$ is the (ignored) mass matrix, $\boldsymbol{F}_{int}$ is the internal forces due to the stresses and $\boldsymbol{F}_{ext}$ is the external loads. This equation will be solved at the end of the time step in our approach. Thus

$$\boldsymbol{0} = {}^{n+1}_{i}\boldsymbol{F}_{\text{ext}} - {}^{n+1}_{i}\boldsymbol{F}_{\text{int}} \qquad [6.14]$$

where the left superscript n is the time increment counter and the subscript i is an iteration counter. These are explained below.

The displacement field is interpolated over an element as

$$ {}^{n+1}_{i}\boldsymbol{u}\left({}^{n+1}_{i}\boldsymbol{x}, t\right) = \boldsymbol{N}_{\text{m}}\left({}^{n+1}_{i}\boldsymbol{x}\right)^{n+1}_{i}\boldsymbol{u}(t) \qquad [6.15]$$

where $\boldsymbol{N}_{\text{m}}({}^{n+1}_{i}\boldsymbol{x})$ is a matrix with the interpolation, shape functions and ${}^{n+1}_{i}\boldsymbol{u}$ is the current estimate of the element displacement vector at the end of the increment and the corresponding coordinate is ${}^{n+1}_{i}\boldsymbol{x}$.

The rate of deformation tensor (velocity strain) is computed as

$$ {}^{n+1}_{i}\boldsymbol{d} = \frac{1}{2}\left(\nabla^{n+1}_{i}\dot{\boldsymbol{u}}^{\text{T}} + \nabla^{n+1}_{i}\dot{\boldsymbol{u}}\right) = {}^{n+1}_{i}\boldsymbol{B}_{\text{m}}\,{}^{n+1}_{i}\dot{\boldsymbol{u}} \qquad [6.16]$$

Note the similarity with the small strain definition. This measure can, as described in Sections 6.3 and 6.4, be used to give the change in stress. The internal forces are obtained from stresses as

$$ {}^{n+1}_{i}\boldsymbol{f}_{\text{int}} = \int_{{}^{n+1}_{i}v^{\text{e}}} {}^{n+1}_{i}\boldsymbol{B}^{\text{T}}_{\text{m}}\,{}^{n+1}_{i}\sigma \text{d}v \xrightarrow{\text{assembling}} {}^{n+1}_{i}\boldsymbol{F}_{\text{int}} \qquad [6.17]$$

The matrix ${}^{n+1}_{i}\boldsymbol{B}$ also relates virtual displacement (velocity) with virtual strain (rate). Thus it has the same appearance as the same matrix used in small deformation analysis to compute strain from displacement. ${}^{n+1}_{i}\sigma$ is the Cauchy stress (true stress) in vector form and ${}^{n+1}_{i}v^{\text{e}}$ is the current volume of the element. ${}^{n+1}_{i}\boldsymbol{F}_{\text{ext}}$ gets contributions from element volume forces as well as from boundary conditions. Note that the load due to the thermal strain is assembled to the internal force vector via the constitutive model. The quasi-static formulation does not need any temporal discretisation. The computation of the stresses is discussed in the next section.

The updating of the current geometry is

$$ {}^{n+1}_{i}\boldsymbol{x} = {}^{n}\boldsymbol{x} + {}_{i}\Delta\boldsymbol{x} = {}^{n}\boldsymbol{x} + {}_{i}\Delta\boldsymbol{U} \qquad [6.18]$$

where the change in coordinate is the incremental displacement of the structure. The iterative procedure will give corrections to this increment as

$$ {}_{i+1}\Delta\boldsymbol{U} = {}_{i}\Delta\boldsymbol{U} + {}_{i}\delta\boldsymbol{U} \qquad [6.19]$$

Equilibrium is set up at the end of the time step/increment and an iteration subscript is introduced. We want to solve

$$ {}^{n+1}_{i}\boldsymbol{R} = {}^{n+1}_{i}\boldsymbol{F}_{\text{ext}} - {}^{n+1}_{i}\boldsymbol{F}_{\text{int}} = \boldsymbol{0} \qquad [6.20]$$

The Newton–Raphson iterative approach is based on a Taylor expansion of this equation where we want to find the iterative correction that may give ${}^{n+1}_{i+1}\boldsymbol{R} = \boldsymbol{0}$ in the next iteration:

$$ {}^{n+1}_{i+1}\boldsymbol{R} \approx {}^{n+1}_{i}\boldsymbol{R} + \frac{\partial {}^{n+1}_{i}\boldsymbol{R}}{\partial \boldsymbol{U}}\, {}_{i}\delta\boldsymbol{U} = \boldsymbol{0} \tag{6.21} $$

This can be written as

$$ -\frac{\partial {}^{n+1}_{i}\boldsymbol{R}}{\partial \boldsymbol{U}}\, {}_{i}\delta\boldsymbol{U} = {}^{n+1}_{i}\boldsymbol{R} \tag{6.22} $$

or introducing the tangent matrix

$$ {}^{n+1}_{i}\boldsymbol{K}_{\mathrm{t}}\, {}_{i}\delta\boldsymbol{U} = {}^{n+1}_{i}\boldsymbol{R} \tag{6.23} $$

where

$$ {}^{n+1}_{i}\boldsymbol{K}_{\mathrm{t}} = -\frac{\partial {}^{n+1}_{i}\boldsymbol{R}}{\partial \boldsymbol{U}} = -\left[\frac{\partial {}^{n+1}_{i}\boldsymbol{F}_{\mathrm{ext}}}{\partial \boldsymbol{U}} - \frac{\partial {}^{n+1}_{i}\boldsymbol{F}_{\mathrm{int}}}{\partial \boldsymbol{U}}\right] \tag{6.24} $$

The dependency of external forces is often ignored, giving

$$ {}^{n+1}_{i}\boldsymbol{K}_{\mathrm{t}} \approx \frac{\partial {}^{n+1}_{i}\boldsymbol{F}_{\mathrm{int}}}{\partial \boldsymbol{U}} \tag{6.25} $$

The internal forces have several contributions that depend on the deformation and they are evaluated element-wise from the integral in Eq. [6.17]. The domain of the integral, the stresses and the ${}^{n+1}_{i}\boldsymbol{B}$-matrix depend on the deformation. The first part is often ignored and the contribution to the tangent stiffness becomes

$$ \begin{aligned} {}^{n+1}_{i}\boldsymbol{k}_{\mathrm{t}} &\approx \frac{\partial {}^{n+1}_{i}\boldsymbol{f}_{\mathrm{int}}}{\partial \boldsymbol{u}} = \frac{\partial}{\partial \boldsymbol{u}}\left[\int_{{}^{n+1}_{i}v^{\mathrm{e}}} {}^{n+1}_{i}\boldsymbol{B}_{\mathrm{m}}^{\mathrm{T}}\, {}^{n+1}_{i}\sigma\, \mathrm{d}v\right] \\ &\approx \int_{{}^{n+1}_{i}v^{\mathrm{e}}} \frac{\partial {}^{n+1}_{i}\boldsymbol{B}_{\mathrm{m}}^{\mathrm{T}}}{\partial \boldsymbol{u}}\, {}^{n+1}_{i}\sigma\, \mathrm{d}v + \int_{{}^{n+1}_{i}v^{\mathrm{e}}} {}^{n+1}_{i}\boldsymbol{B}_{\mathrm{m}}^{\mathrm{T}} \frac{\partial {}^{n+1}_{i}\sigma}{\partial \boldsymbol{u}}\, \mathrm{d}v \end{aligned} \tag{6.26} $$

The first term is the geometric or stress stiffness matrix

$$ {}^{n+1}_{i}\boldsymbol{k}_{\sigma} = \int_{{}^{n+1}_{i}v^{\mathrm{e}}} \frac{\partial {}^{n+1}_{i}\boldsymbol{B}_{\mathrm{m}}^{\mathrm{T}}}{\partial \boldsymbol{u}}\, {}^{n+1}_{i}\sigma\, \mathrm{d}v \tag{6.27} $$

The last term is called the constitutive or material stiffness matrix. It is the contribution from the constitutive matrix as the derivative of the stresses due to the displacement can be written as

$$ \begin{aligned} {}^{n+1}_{i}\boldsymbol{k}_{\mathrm{c}} &= \int_{{}^{n+1}_{i}v^{\mathrm{e}}} {}^{n+1}_{i}\boldsymbol{B}_{\mathrm{m}}^{\mathrm{T}} \frac{\partial {}^{n+1}_{i}\sigma}{\partial \boldsymbol{u}} = \int_{{}^{n+1}_{i}v^{\mathrm{e}}} {}^{n+1}_{i}\boldsymbol{B}_{\mathrm{m}}^{\mathrm{T}} \frac{\partial {}^{n+1}_{i}\sigma}{\partial \varepsilon} \frac{\partial \varepsilon}{\partial \boldsymbol{u}} \\ &= \int_{{}^{n+1}_{i}v^{\mathrm{e}}} {}^{n+1}_{i}\boldsymbol{B}_{\mathrm{m}}^{\mathrm{T}}\, {}^{n+1}_{i}\boldsymbol{c}_{\mathrm{t}}\, {}^{n+1}_{i}\boldsymbol{B}_{\mathrm{m}}\, \mathrm{d}v \end{aligned} \tag{6.28} $$

where the consistent constitutive matrix, or sometimes called algorithmic constitutive matrix, is

$$ {}_{i}^{n+1}\boldsymbol{c}_{\mathrm{t}} = \frac{\partial_i^{n+1}\sigma}{\partial \varepsilon} \qquad [6.29] $$

The last equality in Eq. [6.28] is motivated as follows. The term $\partial_i^{n+1}\sigma/\partial \boldsymbol{u}$ relates perturbation in displacement, $\delta\boldsymbol{u}$ and corresponding perturbation in stress $\delta\sigma$. The last equality in Eq. [6.28] has this replaced by ${}_{i}^{n+1}\boldsymbol{c}_{\mathrm{t}}\,{}^{n+1}\boldsymbol{B}_{\mathrm{m}}$. Thus we have

$$ \delta\boldsymbol{\sigma} = {}_{i}^{n+1}\boldsymbol{c}_{\mathrm{t}}\,{}_{i}^{n+1}\boldsymbol{B}_{\mathrm{m}}\,\delta\boldsymbol{u} \qquad [6.30] $$

The matrix ${}_{i}^{n+1}\boldsymbol{B}_{\mathrm{m}}$ in Eq. [6.16] also holds for

$$ \delta\boldsymbol{\varepsilon} = {}_{i}^{n+1}\boldsymbol{B}_{\mathrm{m}}\,\delta\boldsymbol{u} \qquad [6.31] $$

Inserting Eq. [6.31] into Eq. [6.30] gives

$$ \delta\boldsymbol{\sigma} = {}_{i}^{n+1}\boldsymbol{c}_{\mathrm{t}}\,\delta\boldsymbol{\varepsilon} \qquad [6.32] $$

Thereby the last equality in Eq. [6.28] is motivated. Its evaluation depends on the computational algorithm used for the stress updating algorithm. Therefore, it is called 'algorithmic' constitutive matrix [95, 99, 100]. Thus we get

$$ {}_{i}^{n+1}\boldsymbol{k}_{\sigma} + {}_{i}^{n+1}\boldsymbol{k}_{\mathrm{c}} \xrightarrow{\text{assembling}} {}_{i}^{n+1}\boldsymbol{K}_{\sigma} + {}_{i}^{n+1}\boldsymbol{K}_{\mathrm{c}} = {}_{i}^{n+1}\boldsymbol{K}_{\mathrm{t}} \qquad [6.33] $$

The assembly of element matrices as well as loads is implemented in the function LEL_Mech_Loop in CWM_Lab.

A modified Newton–Raphson approach may also be more efficient in mechanical analysis as stated for the thermal analysis earlier. It is recommended to combine it with line search [92] if no contact surfaces are present in the model. The incremental, iterative approach for the nonlinear deformation problems is summarised in Fig. 6.4.

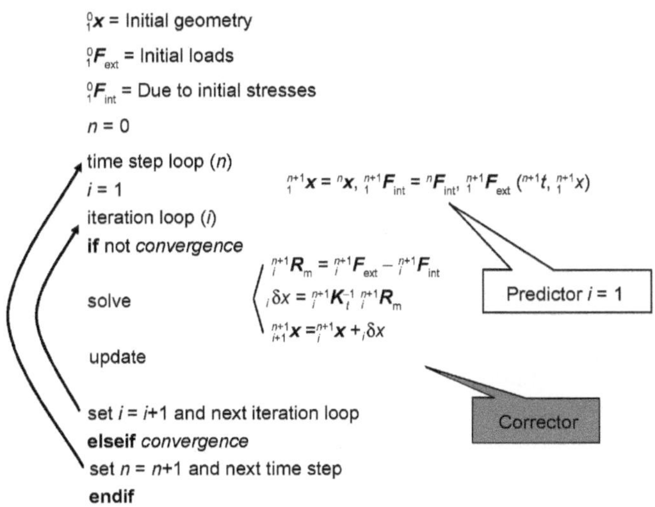

6.4 Incremental, iterative procedure for mechanical analysis implemented in the function LEL_Mechmod in CWM_Lab.

6.3 Constitutive model

The UL formulation is based on the Cauchy (true) stress and the rate of deformation tensor (velocity strain) as conjugate stress and strain measures [34, 93]. The basic purpose of the constitutive model and the corresponding numerical algorithm is to update stresses due to the deformation and also, in the case of implicit finite element codes, provide a consistent constitutive matrix, Eq. [6.29]. The first task must be done efficiently and accurately and is described in Section 6.4. The computation of a consistent tangent is required when the second order convergence of the Newton–Raphson method needs to be preserved. The latter can be compromised as described earlier. The computation of the consistent constitutive matrix depends on the material model and the numerical algorithm used [95, 99]. This subject will not be dealt with here. The stress-updating algorithm will be form-identical [101] with the small strain logic when it is formulated using the concepts and models outlined below. See also Crisfield [93] where several approaches for large strain plasticity leading to the 'conventional' rate form are described. Bonet and Wood [94] showed in their appendix A how a hyperelastoplastic formulation using a logarithmic strain definition in principal stress space also leads to this.

There are two basic variants of how to compute stresses from strains (the left-hand arrow pointing upwards in Fig. 6.1). The most common one is assuming hypoelastic material model behaviour. This relates increment in strain with an objective increment in stress. The hypoelastic models do not exactly correspond to an elastic material but the error disappears in the case of metal plasticity as the elastic strains are small for this case [102]. The use of hypoelastic relations also has the drawback that an objective stress rate measure must be chosen.

The alternative, used in the finite element code CWM_Lab is to use a hyperelastic model (see pages xiii–xiv). Then the deformation is used to compute the deformation gradient and a multiplicative decomposition of the latter into elastic and plastic parts is assumed. The rotation of the plastic part is usually assumed to be zero by motivating from crystal plasticity. The plastic deformation of a crystal lattice due to dislocation glide takes place by a shearing process only without any rotation. The elastic part can be used together with a strain energy definition to compute the total stress directly. Thus the need for objective stress rates is circumvented.

The first approach usually starts by assuming the additive split of different strain rate contributions whereas the latter approach starts with the multiplicative decomposition of the deformation gradient. The use of logarithmic strain definition combined with performing the stress update in the principal stretch directions leads to an additive decomposition of the strain rates of the plastic and elastic parts as in the hypoelastic approach. Both formulations can thus be expressed in a format that can be handled by the radial-return stress update algorithm described in Section 6.4.

6.3.1 Hypoelastoplastic formulation

The starting point can be the assumption about additive decomposition of the different contributions to the strain rate

$$\boldsymbol{d} = \boldsymbol{d}^{\mathrm{e}} + \boldsymbol{d}^{\mathrm{p}} + \boldsymbol{d}^{\mathrm{th}} \qquad [6.34]$$

where $\boldsymbol{d}$ is the total strain rate, $\boldsymbol{d}^{\mathrm{e}}$ is the elastic strain rate, $\boldsymbol{d}^{\mathrm{p}}$ is the plastic strain rate and $\boldsymbol{d}^{\mathrm{th}}$ rate of the thermal strain. The latter is purely volumetric. The additive decomposition can be motivated in different ways starting from the multiplicative decomposition of the deformation gradient (see for example [93, 95, 103–105]). Some of the differences between these derivations disappear when assuming small elastic strains (see chapter 19 in [93] and [102]). This is reasonable for metal plasticity. We will leave out the thermal strain in the following discussion. It is a volumetric strain and does not affect the plasticity calculations as we will deal only with deviatoric plasticity. The effect of temperature on the elastoplastic properties will be included in the description of the stress update algorithm.

The strain increment is obtained from the objective spatial velocity gradient in Eq. [6.16]: compare with the relation used in Eq. [6.31]. Hughes and Winget [106] showed that the formula below is a second order accurate approximation of this increment

$$_{i}\Delta\varepsilon = \int_{^{n}t}^{^{n+1}t} {}_{i}\boldsymbol{d}(\tau)\mathrm{d}\tau \approx {}^{n+1/2}_{i}\boldsymbol{B}_{\mathrm{m}}\, {}_{i}\Delta\boldsymbol{u} \qquad [6.35]$$

where ${}^{n+1/2}_{i}\boldsymbol{B}_{\mathrm{m}}$ means that the matrix with derivatives of the shape functions is evaluated for the current estimate of the midpoint geometry ${}^{n+1/2}_{i}\boldsymbol{x}$. The additive decomposition of the strain rates in Eq. [6.34] gives the corresponding decomposition of the total strain increment.

The basic idea for the hypoelastic formulation [34, 93, 95] is to relate an objective stress rate to strain rate or rather to relate an objective increment in stress with an objective increment in strain. The minimum requirement for objectivity of a tensor is that a superimposed rigid body rotation should give a change in the tensor according to the standard rules for tensor. The spatial velocity gradient tensor and the stress tensor are examples of objective. Therefore the strain increment according to Eq. [6.35] is also objective; however, the time derivative of the stress is not objective.

The introduction of objective stress rates brings in additional terms in order to fulfil this objectivity requirement. The increment in objective stress rate is related to elastic part of the increment strain

$$\boldsymbol{\sigma}^{\nabla} = \boldsymbol{E}\boldsymbol{d}^{\mathrm{e}} \qquad [6.36]$$

where $\boldsymbol{\sigma}^{\nabla}$ is any objective stress rate, $\boldsymbol{E}$ is the matrix with elastic properties, and $\boldsymbol{d}^{\mathrm{e}}$ is the elastic strain rate. (See examples in Section 6.3.3 for the stress updating in the elastic case.) There exist several proposals for objective stress rates and the focus is on the Green–Naghdi stress rate below. It is defined as

$$\boldsymbol{\sigma}^{\nabla GN} = \dot{\boldsymbol{\sigma}} - \dot{\boldsymbol{R}}\boldsymbol{R}^{\mathrm{T}}\boldsymbol{\sigma} + \boldsymbol{\sigma}\dot{\boldsymbol{R}}\boldsymbol{R}^{\mathrm{T}} \qquad [6.37]$$

where $\boldsymbol{R}$ is the rotation tensor obtained from polar decomposition of the deformation gradient. This rate corresponds to the use of co-rotating stress [34, 95, 101]:

$$\boldsymbol{\sigma}_{\mathrm{R}} = \boldsymbol{R}^{\mathrm{T}}\boldsymbol{\sigma}\boldsymbol{R} \qquad [6.38]$$

The use of a co-rotated stress makes it possible to implement the stress updating by performing it in the co-rotated state as shown below. The initial state is rotated as

$$\begin{aligned} {}^{n}\boldsymbol{\sigma}_{\mathrm{R}} &= {}^{n}\boldsymbol{R}^{\mathrm{T}}\,{}^{n}\boldsymbol{\sigma}\,{}^{n}\boldsymbol{R} \\ {}^{n}\boldsymbol{\varepsilon}^{\mathrm{p}}_{\mathrm{R}} &= {}^{n}\boldsymbol{R}^{\mathrm{T}}\,{}^{n}\boldsymbol{\varepsilon}^{\mathrm{p}}\,\boldsymbol{R} \end{aligned} \qquad [6.39]$$

and the same for the midpoint strain increment

$$_{i}\Delta\boldsymbol{\varepsilon}_{\mathrm{R}} = {}^{n+1/2}_{i}\boldsymbol{R}^{\mathrm{T}}\,{}_{i}\Delta\boldsymbol{\varepsilon}\,{}^{n+1/2}_{i}\boldsymbol{R} \qquad [6.40]$$

Thereafter the stress updating is done in this configuration (see Section 6.4). The stress update is done in deviatoric space as the pressure part is a pure thermoelastic term.

The hypoelastic deviatoric stress update using co-rotated stresses becomes:

$$\begin{aligned} {}^{n+1}\boldsymbol{s}_{\mathrm{R}} &= {}^{n+1}G\left({}^{n}\boldsymbol{e}^{\mathrm{e}}_{\mathrm{R}} + \Delta\boldsymbol{e}^{\mathrm{e}}_{\mathrm{R}}\right) = \frac{{}^{n+1}G}{{}^{n}G}\,{}^{n}\boldsymbol{s}_{\mathrm{R}} + 2\,{}^{n+1}G(\Delta\boldsymbol{e}_{\mathrm{R}} - \Delta\boldsymbol{e}^{\mathrm{p}}_{\mathrm{R}}) \\ &= {}^{\mathrm{trial}}\boldsymbol{s}_{\mathrm{R}} - 2\,{}^{n+1}G\Delta\boldsymbol{e}^{\mathrm{p}}_{\mathrm{R}} \end{aligned} \qquad [6.41]$$

The logic for the determination of $\Delta\boldsymbol{e}^{\mathrm{p}}_{\mathrm{R}}$ is given in Section 6.4. Temperature-dependent material properties are accounted for since the left superscript on the material properties denotes that they are evaluated at corresponding temperature. The trial state is introduced, and denoted by ${}^{\mathrm{trial}}[\ ]$. This is the state that would exist at time ${}^{n+1}t$ provided no plastic deformation occurs during the increment. The stress updating is then done for the deviatoric stresses (Eq. [6.41]), and thereafter is the pressure part added to give the total stress tensor. The thermal dilatation is included in this volumetric part. However, temperature effects are still present in the updating of the deviatoric part due to temperature-dependent material elastic and plastic properties. The updated stress state and other variables, such as plastic strain components, are finally rotated to current geometry by

$$\begin{aligned} {}^{n+1}_{i}\boldsymbol{\sigma} &= {}^{n+1}\boldsymbol{R}\,{}^{n+1}_{i}\boldsymbol{\sigma}_{\mathrm{R}}\,{}^{n+1}_{i}\boldsymbol{R}^{\mathrm{T}} \\ {}^{n+1}_{i}\boldsymbol{\varepsilon}^{\mathrm{p}} &= {}^{n+1}_{i}\boldsymbol{R}\,{}^{n+1}_{i}\boldsymbol{\varepsilon}^{\mathrm{p}}_{\mathrm{R}}\,{}^{n+1}_{i}\boldsymbol{R}^{\mathrm{T}} \end{aligned} \qquad [6.42]$$

6.3.2 Hyperelastoplastic formulation

The hyperelastoplastic algorithm is based on computing the current stress directly from the strain energy. There are several variants but the basic idea is that the derivative of the strain energy is taken with respect to an objective strain measure. Therefore, the stress update will be objective without any need to choose so-called objective rates. See examples in Section 6.3.3.

The approach implemented in CWM_Lab is illustrated below. The appendix in Bonet and Wood [94] provide details and Belytschko *et al.* [34] give more general formulations. A formulation starting with the multiplicative decomposition of the deformation gradient leading to an additive split of deformation rates is used. Then the same stress update logic, Section 6.4, can be used for the hypoelastoplastic and hyperelastoplastic approaches.

The strain energy function is assumed to be unchanged by the plastic deformation. The model in CWM_Lab is defined with respect to original volume and therefore the Kirchhoff stress, $\boldsymbol{\tau}$, is obtained when taking the derivative of the strain energy with respect to the strain measure. The relation between Cauchy stress and Kirchhoff stress is

$$\boldsymbol{\sigma} = \frac{\boldsymbol{\tau}}{J} \tag{6.43}$$

where J is the relative volume change obtained from the deformation gradient as

$$J = \det \boldsymbol{F} \tag{6.44}$$

The multiplicative split is written as

$$\mathrm{d}\boldsymbol{x} = \boldsymbol{F}\mathrm{d}\boldsymbol{X} = \boldsymbol{F}_{\mathrm{e}}\boldsymbol{F}_{\mathrm{p}}\mathrm{d}\boldsymbol{X} = J^{-1/3}\,{}_{\mathrm{dev}}\boldsymbol{F}_{\mathrm{e}}\boldsymbol{F}_{\mathrm{p}}\mathrm{d}\boldsymbol{X} \tag{6.45}$$

The volumetric part of the elastic deformation tensor has in the last equality been removed from ${}_{\mathrm{dev}}\boldsymbol{F}_{\mathrm{e}}$ as a prelude to the description of the stress update algorithm in presence of plasticity in Section 6.4. The plastic part of the deformation gradient has no volume change in the case of deviatoric plasticity. A pure thermoelastic update can then be done for the hydrostatic part of the stresses and therefore it need not be included in the stress update algorithm.

The multiplicative split above introduces the notion about a local, intermediate unloaded configuration (Fig. 6.5). The split into elastic and plastic parts of the deformation gradient does not uniquely determine the rotation of this configuration. Either an evolution equation for the plastic spin must be given or, it is assumed to be zero. The latter is usually motivated by crystal plasticity [95], as plastic deformation dominated by dislocation glide will only cause shearing of the crystal lattices and no rotation. Assuming zero plastic spin is a common approach and used in CWM_Lab: all the rigid rotation of the motion is then part of $\boldsymbol{F}_{\mathrm{e}}$. The stress updating will take place in this unloaded configuration in the case of plasticity. Its rotation does not matter in the case of an isotropic material.

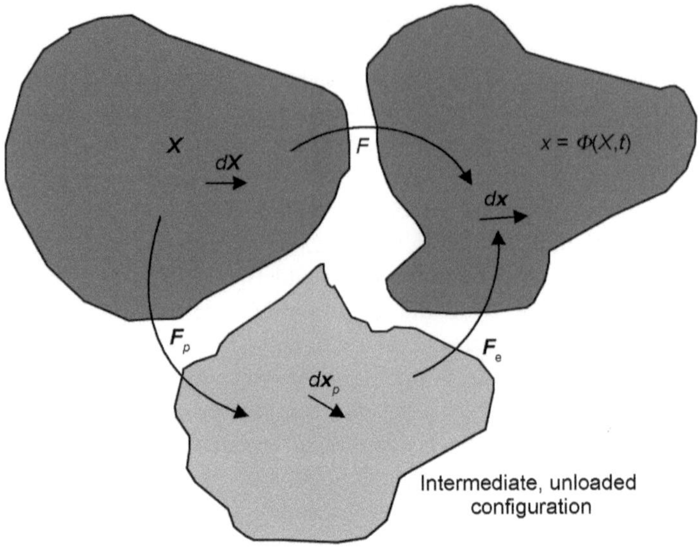

6.5 Multiplicative split of deformation gradient and intermediate, unloaded configuration.

The Finger tensor is the measure of the strain in the current configuration that is used in CWM_Lab. It can be separated into volumetric and deviatoric parts as

$$\boldsymbol{b} = \boldsymbol{F}\boldsymbol{F}^{\mathrm{T}} = J^{-2/3}\,{}_{\mathrm{dev}}\boldsymbol{F}\,{}_{\mathrm{dev}}\boldsymbol{F}^{\mathrm{T}} = J^{-2/3}\,{}_{\mathrm{dev}}\boldsymbol{b} \qquad [6.46]$$

The term ${}_{\mathrm{dev}}\boldsymbol{b}$ is the deviatoric part of the Finger tensor and J gives the volumetric part. There is also a need for a measure of the elastic deformation in the current configuration as this determines the stress. This is, by use of the multiplicative split in Eq. [6.45],

$$\boldsymbol{b}^{\mathrm{e}} = \boldsymbol{F}_{\mathrm{e}}\boldsymbol{F}_{\mathrm{e}}^{\mathrm{T}} = \boldsymbol{F}\boldsymbol{F}_{\mathrm{p}}^{-1}\boldsymbol{F}_{\mathrm{p}}^{-\mathrm{T}}\boldsymbol{F}^{\mathrm{T}} = \boldsymbol{F}\boldsymbol{C}_{\mathrm{p}}^{-1}\boldsymbol{F}^{\mathrm{T}} \qquad [6.47]$$

as $\boldsymbol{F}_{\mathrm{e}}$ relates the unloaded and current configurations (see Fig. 6.5 and Eq. [6.9]), by

$$\mathrm{d}\boldsymbol{x} = \boldsymbol{F}_{\mathrm{e}}\,\mathrm{d}\boldsymbol{x}_{\mathrm{p}} \qquad [6.48]$$

The introduced right Cauchy strain tensor

$$\boldsymbol{C}_{\mathrm{p}} = \boldsymbol{F}_{\mathrm{p}}^{\mathrm{T}}\boldsymbol{F}_{\mathrm{p}} \qquad [6.49]$$

is a measure of the plastic deformation on the unloaded configuration. The latter is used in the implementation in CWM_Lab as the memory of the plastic deformation history. Note that the corresponding deviatoric version ${}_{\mathrm{dev}}\boldsymbol{b}^{\mathrm{e}}$ can be introduced by using ${}_{\mathrm{dev}}\boldsymbol{F}$ in Eq. [6.47] and that the plastic strains are purely deviatoric in our case.

The split into elastic and plastic parts can be visualised for uniaxial deformation by the example in Fig. 6.6. Then the deformation gradient is equal to the stretch in Eq. [6.3] as we have

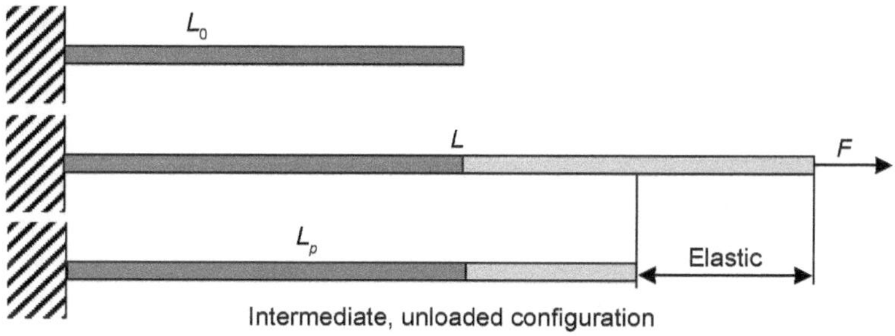

6.6 Original, current and unloaded configurations of rod.

$$dx = F dX = \lambda dX \quad [6.50]$$

Unloading the rod gives a permanent change in length due to the plastic deformation. The multiplicative split of the deformation gradient becomes

$$\lambda = \frac{L}{L_0} = \frac{L}{L_p}\frac{L_p}{L_0} = \lambda^e \lambda^p \quad [6.51]$$

This corresponds to the split in Eq. [6.45]. The elastic stretch can be used to characterise the elastic deformation or the elastic Finger tensor in Eq. [6.47]. The latter becomes

$$b^e = (\lambda^e)^2 = \left(\frac{\lambda}{\lambda^p}\right)^2 \quad [6.52]$$

for this uniaxial example. The last equality is obtained by using Eq. [6.51].

The use of the uniaxial version of the velocity strain in Eq. [6.4] makes it possible to obtain an additive split as:

$$d = \dot{\lambda}\frac{1}{\lambda} = \frac{\dot{\lambda}^e \lambda^p}{\lambda^e \lambda^p} + \frac{\lambda^e \dot{\lambda}^p}{\lambda^e \lambda^p} = \frac{\dot{\lambda}^e}{\lambda^e} + \frac{\dot{\lambda}^p}{\lambda^p} = d^e + d^p \quad [6.53]$$

The stress-updating logic in Section 6.4 is based on an elastic predictor phase followed by a plastic corrector step. This means that it is first assumed that the plastic increment is zero. This is the trial state. The trial elastic Finger tensor, Eq. [6.47], becomes

$$^{\text{trial}}\boldsymbol{b}^e = {}^{n+1}\boldsymbol{F}\,{}^{n}\boldsymbol{C}_p^{-1}\,{}^{n+1}\boldsymbol{F}^T \quad [6.54]$$

and the trial stretch in the uniaxial case is

$$^{\text{trial}}\lambda^e = \frac{^{n+1}\lambda^e}{^{n}\lambda^p} = {}^{\nabla}\lambda\,{}^{n}\lambda^e \quad [6.55]$$

where ${}^{\nabla}\lambda$ is the contribution to the stretch by the current increase in the displacement. It is assumed to be completely elastic in the trial state. Taking the logarithm of the last equation gives

$$\ln\left(^{\text{trial}}\lambda^e\right) = \ln(^{n}\lambda^e) + \ln\left(^{\nabla}\lambda\right) \quad [6.56]$$

This can be interpreted as a relation between the strains when using the logarithmic strain definition in Eq. [6.2]. The right-hand side above then corresponds to the term ${}^{n}\boldsymbol{e}^{\mathrm{e}}_{\mathrm{R}} + \Delta\boldsymbol{e}^{\mathrm{e}}_{\mathrm{R}}$ in Eq. [6.41] for the hypoelastic approach. The general tensor notations have been exemplified by corresponding uni-axial notations above. Equation [6.56] shows that the multiplicative split of the deformation gradient leads to an additive split in strain rates that is parallel to the hypoelastic approach in the previous chapter. A generalisation of Eq. [6.56] to general stress states can only be done by use of principal elastic stretches where the logarithm can be applied to give corresponding principal elastic strains. Then Eq. [6.56] is written as

$$\ln\left({}^{\mathrm{trial}}_{\mathrm{dev}}\lambda^{\mathrm{e}}_{\alpha}\right) = \ln\left({}^{n}\lambda^{\mathrm{e}}_{\alpha}\right) + \ln\left({}^{\nabla}\lambda_{\alpha}\right), \qquad \alpha = 1\ldots3 \qquad [6.57]$$

The stress update for deviatoric plasticity discussed in this book is conveniently performed in deviatoric stress space. Trial deviatoric stresses are computed first and a plastic correction will be done in case of yielding as in the description of the hypoelastic model in the previous section. Thus we will use the principal deviatoric trial elastic stretches

$$\ln\left({}^{\mathrm{trial}}_{\mathrm{dev}}\lambda^{\mathrm{e}}_{\alpha}\right) = \ln\left({}^{n}_{\mathrm{dev}}\lambda^{\mathrm{e}}_{\alpha}\right) + \ln\left({}^{\nabla}_{\mathrm{dev}}\lambda_{\alpha}\right), \qquad \alpha = 1\ldots3 \qquad [6.58]$$

A strain energy function where the volumetric part and the deviatoric parts are separated is used. The trial principal deviatoric stresses are then obtained by

$${}^{\mathrm{trial}}_{\mathrm{dev}}\tau_{\alpha} = \frac{\partial W}{\partial\, {}^{\mathrm{trial}}_{\mathrm{dev}}\lambda'^{\mathrm{e}}_{\alpha}}, \qquad \alpha = 1\ldots3 \qquad [6.59]$$

The subsequent plastic corrector phase will then need a reduction of this stress as part of the total strain increment is plastic. The term $\ln\left({}^{\nabla}_{\mathrm{dev}}\lambda_{\alpha}\right)$ in Eq. [6.58] is the total strain increment. The plastic part of this strain increment is denoted $\ln\left({}^{\nabla}\lambda^{\mathrm{p}}_{\alpha\alpha}\right)$. The logic for its determination is given in Section 6.4. The plastic corrector formula is in the current hyperelasto-plastic case:

$${}^{n+1}_{\mathrm{dev}}\tau_{\alpha\alpha} = {}^{\mathrm{trial}}_{\mathrm{dev}}\tau_{\alpha\alpha} - 2\,{}^{n+1}G\ln\left({}^{\nabla}\lambda^{\mathrm{p}}_{\alpha\alpha}\right), \qquad \alpha = 1\ldots3 \qquad [6.60]$$

It corresponds to the last equality in Eq. [6.41] for the hypoelastoplastic model.

6.3.3 Comparison of hypoelastic and hyperelastic formulations

The purpose of this section is to give an example where the concepts in Section 6.3.1 and 6.3.2 are shown at work. Furthermore, the pure shear case has been extensively discussed in the literature [34, 93, 107] and has affected the implementation of hypoelastoplastic models in many commercial finite element programs. This example, together with the realisation that the hypoelastic formulation does not define an elastic material in the general case [102] (although the error is negligible in the case of small elastic strains), has paved the way for the implementation of hyperelastoplastic models. This case was

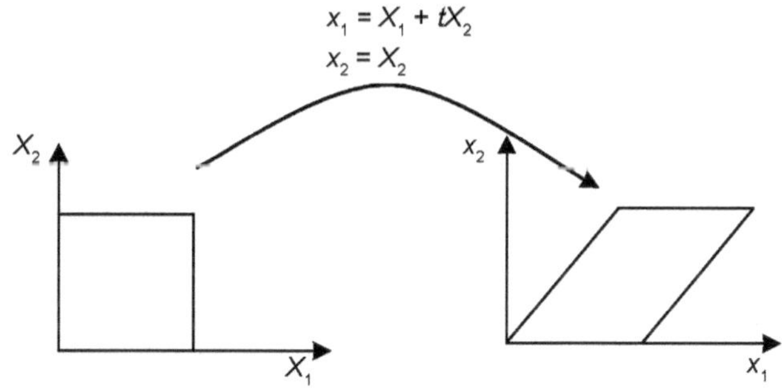

6.7 Pure shear deformation.

investigated by [108, 109] for elasto-perfectly plastic case for Jaumann, Green–Naghdi and Truesdell stress rates.

The case of pure shear (see Fig. 6.7) is described mathematically as a mapping from the original geometry, $\boldsymbol{X}$, to current configuration, $\boldsymbol{x}$, by

$$x_1 = X_1 + tX_2$$
$$x_2 = X_2 \qquad [6.61]$$

Equation [6.47] is used to compute the deformation gradient, which becomes

$$\boldsymbol{F} = \frac{\partial \boldsymbol{x}}{\partial \boldsymbol{X}} = \begin{bmatrix} 1 & t \\ 0 & 1 \end{bmatrix} \qquad [6.62]$$

The following tensors are of interest

$$\boldsymbol{F}^{-1} = \begin{bmatrix} 1 & t \\ 0 & 1 \end{bmatrix} \qquad [6.63]$$

and

$$\dot{\boldsymbol{F}} = \begin{bmatrix} 0 & 1 \\ 0 & 0 \end{bmatrix} \qquad [6.64]$$

Different deformation measures can now be computed. The left Cauchy–Green or Finger tensor is defined as:

$$\boldsymbol{b} = \boldsymbol{F}\boldsymbol{F}^{\mathrm{T}} = \begin{bmatrix} 1+t^2 & t \\ t & 1 \end{bmatrix} \qquad [6.65]$$

This tensor described the deformation in the current configuration whereas the right Cauchy–Green tensor

$$\boldsymbol{C} = \boldsymbol{F}^{\mathrm{T}}\boldsymbol{F} = \begin{bmatrix} 1 & t \\ t & 1+t^2 \end{bmatrix} \qquad [6.66]$$

describes the deformation in the original configuration.

The eigenvalues of $\boldsymbol{b}$ give the principal stretches in square. The principal stretches are

$$\lambda_1 = \frac{1 + \sqrt{t^2 + 4}}{2} \qquad \text{and} \qquad \lambda_2 = \frac{-1 + \sqrt{t^2 + 4}}{2} \tag{6.67}$$

or more conveniently

$$\lambda_1 = \frac{1 + s}{c} \qquad \text{and} \qquad \lambda_2 = \frac{1 - s}{c} \tag{6.68}$$

with

$$s = \sin(\beta), c = \cos(\beta) \text{ and } t = 2\tan(\beta) \tag{6.69}$$

These stretches are invariant. Thus the same results would be obtained if the right Cauchy–Green tensor is used. The (orthogonal) eigenvectors for $\boldsymbol{b}$, are the principal stretch directions in current configuration

$$\boldsymbol{n}_1 = \frac{1}{\sqrt{2(1+s)}} \begin{bmatrix} 1+s \\ c \end{bmatrix} \qquad \text{and} \qquad \boldsymbol{n}_2 = \frac{1}{\sqrt{2(1+s)}} \begin{bmatrix} -c \\ 1+s \end{bmatrix} \tag{6.70}$$

The (right) polar decomposition of the deformation gradient is defined as

$$\boldsymbol{F} = \boldsymbol{R}\boldsymbol{U} \tag{6.71}$$

The rotation matrix $\boldsymbol{R}$ can be obtained via the polar decomposition theorem

$$\boldsymbol{R} = \begin{bmatrix} c & s \\ -s & c \end{bmatrix} \tag{6.72}$$

and its rate is needed. We will use the relation

$$\beta = \tan^{-1}\left(\frac{t}{2}\right) \rightarrow \frac{\mathrm{d}\beta}{\mathrm{d}t} = \frac{2}{4 + t^2} \tag{6.73}$$

This is obtained by use of Eq. [6.54]. Then the rate of the rotation tensor is

$$\dot{\boldsymbol{R}} = \begin{bmatrix} -s & c \\ -c & -s \end{bmatrix} \frac{\mathrm{d}\beta}{\mathrm{d}t} = \begin{bmatrix} -s & c \\ -c & -s \end{bmatrix} \frac{2}{4 + t^2} \tag{6.74}$$

The principal stretches can also be obtained from an eigenvalue analysis $\boldsymbol{C}$ but the principal directions $\boldsymbol{N}_\alpha$ will then be different as the right Cauchy–Green tensor is associated with the original configuration. The relations between the principal directions are

$$\boldsymbol{n}_\alpha = \boldsymbol{R}\boldsymbol{N}_\alpha \qquad \text{where } \alpha = 1, 2 \tag{6.75}$$

The spatial velocity gradient becomes

$$\boldsymbol{l} = \dot{\boldsymbol{F}}\boldsymbol{F}^{-1} = \begin{bmatrix} 0 & 1 \\ 0 & 0 \end{bmatrix} \tag{6.76}$$

The rate of deformation tensor, velocity strain, is

$$\boldsymbol{d} = \frac{1}{2}\left(\boldsymbol{l} + \boldsymbol{l}^{\mathrm{T}}\right) = \frac{1}{2}\begin{bmatrix} 0 & 1 \\ 1 & 0 \end{bmatrix} \tag{6.77}$$

and the spin tensor

$$\boldsymbol{w} = \frac{1}{2}\left(\boldsymbol{l} + \boldsymbol{l}^{\mathrm{T}}\right) = \frac{1}{2}\begin{bmatrix} 0 & 1 \\ -1 & 0 \end{bmatrix} \qquad [6.78]$$

is needed.

The elastic constitutive matrix is denoted $\boldsymbol{E}$. It can be expressed in terms of the Lame constants, giving pressure and shear contributions

$$\boldsymbol{E}\boldsymbol{d} = \lambda tr(\boldsymbol{d})\boldsymbol{I} + 2G\boldsymbol{d}_{\mathrm{dev}} = 2G\boldsymbol{d} \qquad [6.79]$$

where the last equality is due to the fact that $\boldsymbol{d}$ is purely deviatoric in the pure shear case studied here. Note that λ is the one of the Lame's constants and not a stretch.

The hypoelastic models will use

$$\boldsymbol{\sigma}^{\nabla} = \dot{\boldsymbol{\sigma}} + \ldots = 2G^{\nabla}\boldsymbol{d} \qquad [6.80]$$

where ... denotes different correction terms in the objective rates. The last equality written as

$$\dot{\boldsymbol{\sigma}} = 2G^{\nabla}\boldsymbol{d} - \ldots \qquad [6.81]$$

which must be solved. The hyperelastic models will compute the stresses without needing to solve this kind of equation. The right superscript on the shear modulus denotes the need for different values when using different models.

The above basic relations will be used for obtaining the stresses for the pure shear deformation case in Fig. 6.7. The cases are primarily taken from Belytschko [34], Crisfield [93] and Bonet and Wood [94]. The hypoelastic approach using three different definitions of objective stress rates (Boxes 6.1 to 6.3), and the hyperelastic approach (Boxes 6.4 and 6.5), are compared in Box 6.6. This comparison is based on using the same shear modulus for all models.

Box 6.1 Hypoelastic model with Jaumann stress rate

Inserting the definition of the Jaumann rate into Eq. [6.80] gives

$$\boldsymbol{\sigma}^{\nabla J} = \dot{\boldsymbol{\sigma}} - \boldsymbol{w}\boldsymbol{\sigma} - \boldsymbol{\sigma}\boldsymbol{w}^{\mathrm{T}} = 2G^{J}\boldsymbol{d} \qquad [6.82]$$

where the spin tensor is given in Eq. [6.78]. Then Eq. [6.81] becomes

$$\dot{\boldsymbol{\sigma}} = 2G^{J}\boldsymbol{d} + \boldsymbol{w}\boldsymbol{\sigma} + \boldsymbol{\sigma}\boldsymbol{w}^{\mathrm{T}} \qquad [6.83]$$

or written explicitly

$$\begin{bmatrix} \dot{\sigma}_{xx} & \dot{\sigma}_{xy} \\ \dot{\sigma}_{xy} & \dot{\sigma}_{yy} \end{bmatrix} = G^{J}\begin{bmatrix} 0 & 1 \\ 1 & 0 \end{bmatrix} + \frac{1}{2}\begin{bmatrix} 0 & 1 \\ -1 & 0 \end{bmatrix}\begin{bmatrix} \sigma_{xx} & \sigma_{xy} \\ \sigma_{xy} & \sigma_{yy} \end{bmatrix} + \frac{1}{2}\begin{bmatrix} \sigma_{xx} & \sigma_{xy} \\ \sigma_{xy} & \sigma_{yy} \end{bmatrix}\begin{bmatrix} 0 & -1 \\ 1 & 0 \end{bmatrix} \qquad [6.84]$$

This equation can be solved to give

$$\begin{bmatrix} \sigma_{xx} & \sigma_{xy} \\ \sigma_{xy} & \sigma_{yy} \end{bmatrix} = G^J \begin{bmatrix} 1-\cos(t) & \sin(t) \\ \sin(t) & \cos(t)-1 \end{bmatrix} \quad [6.85]$$

Box 6.2 Hypoelastic model with Truesdell stress rate

Inserting the definition of the Truesdell rate into Eq. [6.80] gives

$$\boldsymbol{\sigma}^{\nabla \mathrm{T}} = \dot{\boldsymbol{\sigma}} + tr(\boldsymbol{d})\boldsymbol{\sigma} - \boldsymbol{l}\boldsymbol{\sigma} - \boldsymbol{\sigma}\boldsymbol{l}^{\mathrm{T}} = 2G^{\mathrm{T}}\boldsymbol{d} \quad [6.86]$$

where the spatial velocity gradient is given in Eq. [6.76]. Then Eq. [6.81] becomes

$$\dot{\boldsymbol{\sigma}} = 2G^{\mathrm{T}}\boldsymbol{d} - tr(\boldsymbol{d})\boldsymbol{\sigma} + \boldsymbol{l}\boldsymbol{\sigma} + \boldsymbol{\sigma}\boldsymbol{l}^{\mathrm{T}} \quad [6.87]$$

or written explicitly

$$\begin{bmatrix} \dot{\sigma}_{xx} & \dot{\sigma}_{xy} \\ \dot{\sigma}_{xy} & \dot{\sigma}_{yy} \end{bmatrix} = G^T \begin{bmatrix} 0 & 1 \\ 1 & 0 \end{bmatrix} + \begin{bmatrix} 0 & 1 \\ 0 & 0 \end{bmatrix} \begin{bmatrix} \sigma_{xx} & \sigma_{xy} \\ \sigma_{xy} & \sigma_{yy} \end{bmatrix} + \begin{bmatrix} \sigma_{xx} & \sigma_{xy} \\ \sigma_{xy} & \sigma_{yy} \end{bmatrix} \begin{bmatrix} 0 & 0 \\ 1 & 0 \end{bmatrix} \quad [6.88]$$

This equation can be solved to give

$$\begin{bmatrix} \sigma_{xx} & \sigma_{xy} \\ \sigma_{xy} & \sigma_{yy} \end{bmatrix} = G^J \begin{bmatrix} t^2 & t \\ t & 0 \end{bmatrix} \quad [6.89]$$

Box 6.3 Hypoelastic model with Green–Naghdi stress rate

Inserting the definition of the Green-Naghdi into Eq. [6.80] gives

$$\boldsymbol{\sigma}^{\nabla GN} = \dot{\boldsymbol{\sigma}} - \dot{\boldsymbol{R}}\boldsymbol{R}^{\mathrm{T}}\boldsymbol{\sigma} - \boldsymbol{\sigma}\boldsymbol{R}\dot{\boldsymbol{R}}^{\mathrm{T}} = 2G^{GN}\boldsymbol{d} \quad [6.90]$$

where the rotation tensor is given in Eq. [6.72] and its rate in Eq. [6.74]. The Eq. [6.81] becomes

$$\dot{\boldsymbol{\sigma}} = 2G^{GN}\boldsymbol{d} + \dot{\boldsymbol{R}}\boldsymbol{R}^{\mathrm{T}}\boldsymbol{\sigma} + \boldsymbol{\sigma}\boldsymbol{R}\dot{\boldsymbol{R}}^{\mathrm{T}} \quad [6.91]$$

or written explicitly, after some manipulations

$$\begin{bmatrix} \dot{\sigma}_{xx} & \dot{\sigma}_{xy} \\ \dot{\sigma}_{xy} & \dot{\sigma}_{yy} \end{bmatrix} = G^{GN} \begin{bmatrix} 0 & 1 \\ 1 & 0 \end{bmatrix} + \frac{2}{4+t^2} \begin{bmatrix} 2\sigma_{xy} & \sigma_{yy}-\sigma_{xx} \\ \sigma_{yy}-\sigma_{xx} & -2\sigma_{xy} \end{bmatrix} \quad [6.92]$$

This equation can be solved [110] to give

$$\begin{aligned} \sigma_{xx} &= -\sigma_{yy} = 4G^{GN}\{\cos(2\beta)\ln[\cos(\beta)] + \beta\sin(2\beta) - \sin^2(\beta)\} \\ \sigma_{xy} &= 2G^{GN}\cos(2\beta)\{2\beta - 2\tan(2\beta)\ln[\cos(\beta)] - \tan(\beta)\} \end{aligned} \quad [6.93]$$

Box 6.4 Hyperelastic compressible Neo–Hookean material model
The strain energy is

$$W = \frac{G^{NH}}{2}(tr(\boldsymbol{C}) - 3) - G^{NH}\ln J + \frac{\lambda^{NH}}{2}(\ln J)^2 \quad [6.94]$$

where the pure shear case has $J = 1$. The second Piola–Kirchhoff stress is computed as

$$\boldsymbol{S} = 2\frac{\partial W}{\partial \boldsymbol{C}} = G^{NH}(\boldsymbol{I} - \boldsymbol{C}^{-1}) + \lambda^{NH}(\ln J)\,\boldsymbol{C}^{-1} \quad [6.95]$$

A push forward [34, 94] gives the Cauchy stress as

$$\boldsymbol{\sigma} = \frac{G^{NH}}{J}(\boldsymbol{b} - \boldsymbol{I}) + \frac{\lambda^{NH}}{J}(\ln J)\,\boldsymbol{I} \quad [6.96]$$

The Finger tensor is given in Eq. [6.65] and the result becomes

$$\boldsymbol{\sigma} = \frac{G^{NH}}{J}\begin{bmatrix} t^2 & t \\ t & 0 \end{bmatrix} \quad [6.97]$$

Box 6.5 Hyperelastic Hencky material model
The Hencky model in [93] uses Kirchhoff stresses. They are, in the current case, the same as Cauchy stresses as there is no volume change, $J = 1$.

The strain energy is

$$W = \frac{G^H}{2}\sum_{\alpha}[\ln(\lambda_\alpha)]^2 + \frac{\lambda^H}{2}(\ln J)^2 \quad [6.98]$$

where the principal stretches are given in Eqs [6.68] and [6.69]. The principal stresses become

$$\boldsymbol{\sigma}_\alpha = 2G^{NH}\ln(\lambda_\alpha) \quad [6.99]$$

These are rotated to the global coordinate system, giving

$$\boldsymbol{\sigma} = 2G^{NH}\sum_{\alpha}\ln(\lambda_\alpha)\boldsymbol{n}_\alpha\boldsymbol{n}_\alpha^{\mathrm{T}} = 2G^{NH}\ln(\lambda_1)\begin{bmatrix} s & c \\ c & -s \end{bmatrix} \quad [6.100]$$

where we have Eq. [6.68]

$$\lambda_1 = \frac{1 + \sin(\beta)}{\cos(\beta)}$$

Box 6.6 Comparisons of models

The shear stress versus shearing is plotted in Fig. 6.8 for the cases in the boxes above using a unit shear modulus for all cases.

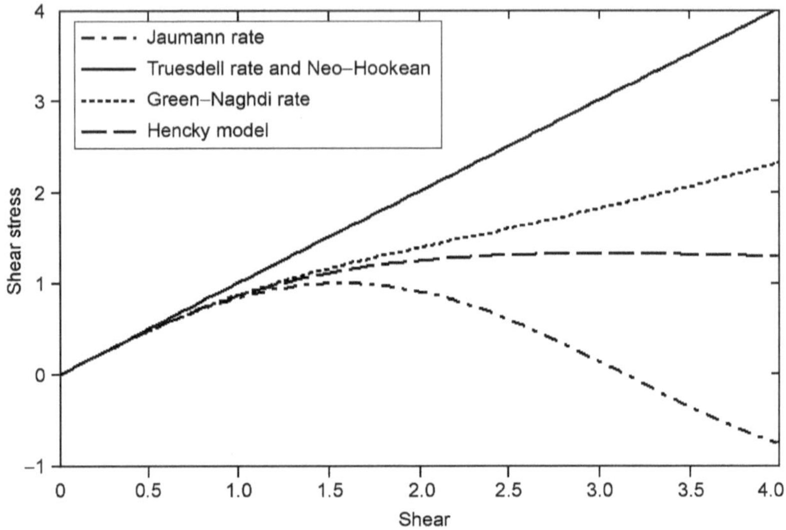

6.8 Shear stress for different models in case of pure shear.

6.4 Stress updating algorithm for deviatoric plasticity

It is assumed that the material is isotropic and the plasticity is deviatoric. Thus it is incompressible and pressure does not affect the plastic behaviour. Furthermore, the associated flow rule is assumed together with isotropic hardening. Introduction of kinematic hardening and pertaining backstress evolution brings with it additional considerations, as the backstress may not be coaxial with the increment in plastic strains. This can be solved by the simultaneous solution of all stresses and back stresses using a fully implicit backward Euler method, described in Section 5.9.2 in [34], or a semi-implicit, described in Section 5.9.6 in [34] or in [111]. A generalised midpoint scheme is used by Chaboche and Cailletaud [112], who discussed different simplifications. Saleeb *et al.* [113] used implicit integration of the system of unknowns, which they found was stable and effective for large increments.

The notations below are given in terms of small strain but the logic is perfectly applicable in the case of large strain plasticity as these approaches, as discussed earlier, can lead to the same kind of incremental formulation. Thus the stresses, $\boldsymbol{s}$, below can be co-rotated deviatoric stresses, $\boldsymbol{s}_R$, if Eq. [6.41] is used.

They correspond to the deviatoric Kirchhoff principal stresses, ${}_{\text{dev}}\tau$, if the hyperelasto-plastic model in Eq. [6.60] is used. The deviatoric strain, $\boldsymbol{e}$, may be rotated deviatoric strains or logarithmic strain from the principal deviatoric stretches, respectively.

The theory of plasticity is summarised first. These relations are given in rate form. Three fundamental relations are needed:

- A condition for when plastic flow occurs.
- A flow rule that determines how the plastic strain develops.
- A hardening rule for the evolution of the yield surface.

The implementation of the radial-return logic is outlined in the following in a manner that can be applied to two types of plasticity models. The first is a model with a yield surface on which the stress state must stay during the plastic deformation. This is usually called rate-independent plasticity. However, it can accommodate rate dependency via the use of a rate-dependent yield limit. The other type of model also uses a yield surface but the stress state can be outside this surface. In the following, the first type of models is denoted plasticity models and the second viscoplastic models.

The surface, f, is used to define elastic and plastic stress processes. A stress state inside, $f \leq 0$, or on the surface $f = 0$ with $\dot{f} \leq 0$, corresponds to an elastic deformation process. The surface is defined as

$$f = \bar{\sigma} - \sigma^{y} \qquad [6.102]$$

where σ^{y} is the yield limit and $\bar{\sigma}$ is the effective von Mises stress

$$\bar{\sigma} = \sqrt{\frac{3}{2}}\,\|\boldsymbol{s}\| \qquad [6.103]$$

where $\|\boldsymbol{s}\|$ is the norm of the deviatoric stress tensor, $\boldsymbol{s}$; see also extensions in [34, 93, 95, 114] to non-isotropic yield surfaces and plane stress.

Furthermore, maximum plastic dissipation is postulated, leading to the associated flow rule [95]. The yield surface is the plastic potential for the plastic strains. It gives the components of the rate of the plastic strain as

$$\dot{\boldsymbol{e}}^{\text{p}} = \dot{\lambda}\,\frac{\partial f}{\partial \boldsymbol{\sigma}} = \dot{\bar{\epsilon}}^{\text{p}}\,\frac{3}{2\bar{\sigma}}\,\boldsymbol{s} \qquad [6.104]$$

The equality $\dot{\lambda} = \dot{\bar{\epsilon}}^{\text{p}}$ can be derived from the relations above. Some different, but equivalent, expressions for the yield surface f, and thereby for Eq. [6.104], are used in the literature. It can be seen that the plastic strains are deviatoric as they are proportional to the deviatoric stress. The effective plastic strain rate $\dot{\bar{\epsilon}}^{\text{p}}$ is used as a measure of the magnitude of the plastic strain rates. It is

$$\dot{\bar{\epsilon}}^{\text{p}} = \sqrt{\frac{2}{3}}\,\|\dot{\boldsymbol{\varepsilon}}^{\text{p}}\| \qquad [6.105]$$

The magnitude of the effective plastic strain rate is determined from the so-called consistency condition in the plasticity model, i.e. the stress state must be on the yield surface during plastic deformation. This consistency condition is

$$f \equiv 0 \qquad [6.106]$$

The magnitude of the effective plastic strain rate is obtained from the flow strength equation in the case of viscoplastic models. Many relations gives the effective plastic strain rate as a monotonous increasing function of the excess stress [115] or overstress where we can have $f > 0$ for this case

$$\dot{\bar{\epsilon}}^{\mathrm{p}} = g\langle f \rangle \qquad [6.107]$$

where the bracket $\langle\ \rangle$ denotes that the rate is zero if $f \leq 0$. Equation [6.107] is one type of flow strength equation. The papers in [115–117] give a number of relations and physical processes such as dislocation glide and diffusion processes, leading to the specific relations in different stress–temperature domains. Frost and Ashby have some of these discussions [118, 119] and several chapters in the book edited by Miller [120] describe processes and models for creep and viscoplasticity. Finally, a hardening rule is needed. The limitation to isotropic hardening means that it is sufficient to define the yield limit as a function of the plastic deformation, i.e. effective plastic strain.

The radial return stress update algorithm is an operator split approach where an elastic predictor is followed by a plastic corrector as indicated by Eqs [6.41] and [6.60]. The two types of models summarised above can be solved by the same approach [100, 112]. The ideas in Section 15.12 in [93] and Section 5.9.8 in [34] and in [111, 119, 121] are followed here. A logarithmic transformation of the flow strength equation can be applied if it is so stiff that the Newton–Raphson scheme cannot find the increment in effective plastic strain [122].

The first step is to compute the trial state. Thereafter, the yield condition is evaluated as

$${}^{\mathrm{trial}}f = {}^{\mathrm{trial}}\bar{\sigma} - {}^{\mathrm{trial}}\sigma^{\mathrm{y}} \qquad [6.108]$$

where the trial yield limit is computed from given data in a format such as

$${}^{\mathrm{trial}}\sigma^{\mathrm{y}} = \sigma^{\mathrm{y}}\left({}^{n+1}T,\ {}^{n}\bar{\epsilon}^{\mathrm{p}}\right) \qquad [6.109]$$

and the trial effective stress is obtained by inserting the trial deviatoric stress into Eq. [6.103]. If the trial state is outside the yield surface, ${}^{\mathrm{trial}}f > 0$, then a plastic corrector must be done. Otherwise the trial state is the true state for ${}^{n+1}t$.

The formula for the plastic corrector is

$${}^{n+1}\boldsymbol{s} = {}^{\mathrm{trial}}\boldsymbol{s} - 2^{n+1}G\Delta\boldsymbol{e}^{\mathrm{p}} \qquad [6.110]$$

This formula corresponds to Eq. [6.41] for the hypoelastoplastic model and Eq. [6.60] for the hyperelastoplastic one. Thus the basic task for the stress update algorithm is to find the plastic strain increment, $\Delta\boldsymbol{e}^{\mathrm{p}}$. The basic assumption in

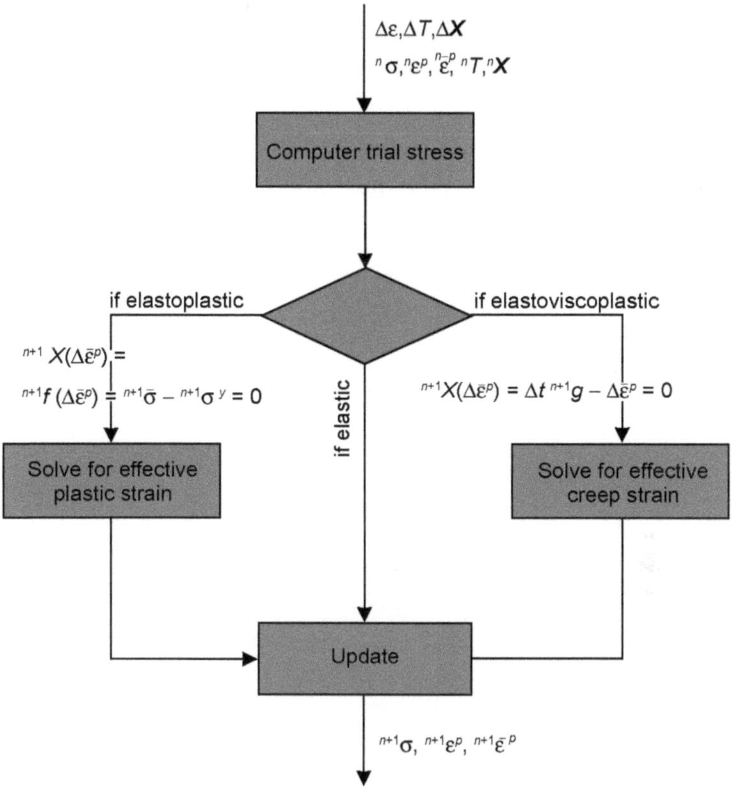

6.9 Overall logic for stress update algorithm for plastic and viscoplastic models.

the radial return approach is to assume that the deviatoric trial stress is used in Eq. [6.104], giving

$$\Delta \boldsymbol{e}^{\mathrm{p}} = \Delta \bar{\epsilon}^{\mathrm{p}} \frac{3}{2^{\mathrm{trial}}\bar{\sigma}} {}^{\mathrm{trial}}\boldsymbol{s} \qquad [6.111]$$

Thus the amount of plastic strain, $\Delta\bar{\epsilon}^{\mathrm{p}}$, remains to be determined. It is obtained from the solution of a nonlinear equation

$$X(\Delta\bar{\epsilon}^{\mathrm{p}}) = 0 \qquad [6.112]$$

where the function is given in Box 6.7. The logic is summarised in Fig. 6.9. Details about the solution of Eq. [6.112] are given in Box 6.7. The derivatives needed for the iterations in this logic are given in Box 6.8 with some further details in Box 6.9.

This radial return is implemented in the CWM_Lab function LEL_RadialReturn for a plasticity model. The model also includes transformation-induced plasticity and effect of changes in phase fractions, $\boldsymbol{X}$, and temperature, T, on material properties (see Sections 9.7.2 and 9.9). The algorithm is combined with a sub-incrementation splitting the increments in temperature, total strain and phase fractions, i.e. the driving variables, into subparts during martensite formation in order to follow the rapid changes in stress due to the transformation-induced plasticity.

Box 6.7 Computation of increment in effective plastic strain

$$X(\Delta\bar{\epsilon}^{\mathrm{p}}) = \begin{cases} {}^{n+1}f & \text{for plasticity} \\ \Delta t^{n+1} g(\langle f \rangle) - \Delta\bar{\epsilon}^{\mathrm{p}} & \text{for viscoplasticity} \end{cases} \quad [6.113]$$

where the first is the consistency equation, Eq. [6.106], and the second is the flow strength equation, Eq. [6.107].

Newton iterations will be used to solve $X(\Delta\bar{\epsilon}^{\mathrm{p}}) = 0$, Eq. [6.112].

1. *Initialise* $i = 1$

$$_1\Delta\bar{\epsilon}^{\mathrm{p}} = 0 \quad [6.114]$$

$$_1X = X(_1\Delta\bar{\epsilon}^{\mathrm{p}}) \quad [6.115]$$

2. *Next estimate* $i = i + 1$

$$_{i+1}\Delta\bar{\epsilon}^{\mathrm{p}} = {}_i\Delta\bar{\epsilon}^{\mathrm{p}} + {}_i\delta\bar{\epsilon}^{\mathrm{p}} \quad [6.116]$$

where a Taylor expansion gives

$$_i\delta\bar{\epsilon}^{\mathrm{p}} = -\frac{_iX}{\left.\dfrac{\mathrm{d}X}{\mathrm{d}\bar{\epsilon}^{\mathrm{p}}}\right|_{_i\Delta\bar{\epsilon}^{\mathrm{p}}}} \quad [6.117]$$

See Box 6.8 for $\dfrac{\mathrm{d}X}{\mathrm{d}\bar{\epsilon}^{\mathrm{p}}}$ for the evaluation of derivatives needed.

3. *Check convergence*. If the function X and/or $_i\delta\bar{\epsilon}^{\mathrm{p}}$ is small enough then stop the iterations, otherwise go back to step 2. Note that X is not nondimensional for the plasticity model and therefore an appropriate scaling should be applied when setting the convergence tolerance.

Box 6.8 Determination of $\dfrac{\mathrm{d}X}{\mathrm{d}\bar{\epsilon}^{\mathrm{p}}}$

X is the yield function f in Eq. [6.113] for plasticity

$$\frac{\mathrm{d}X(\Delta\bar{\epsilon}^{\mathrm{p}})}{\mathrm{d}\bar{\epsilon}^{\mathrm{p}}} = \frac{\mathrm{d}^{n+1}f}{\mathrm{d}\bar{\epsilon}^{\mathrm{p}}} = \frac{\mathrm{d}^{n+1}\bar{\sigma}}{\mathrm{d}\bar{\epsilon}^{\mathrm{p}}} - \frac{\mathrm{d}^{n+1}\sigma_{\mathrm{y}}}{\mathrm{d}\bar{\epsilon}^{\mathrm{p}}} = -\left(3^{n+1}G + \frac{\mathrm{d}\sigma^{\mathrm{y}}}{\mathrm{d}\bar{\epsilon}^{\mathrm{p}}}\right) \quad [6.118]$$

The derivative $\dfrac{d^{n+1}\bar{\sigma}}{d\bar{\epsilon}^{\mathrm{p}}}$ is derived in Box 6.9, Eq. [6.124], and the derivative of the yield limit with respect to effective plastic strain is

$$\frac{\mathrm{d}\sigma^{\mathrm{y}}}{\mathrm{d}\bar{\epsilon}^{\mathrm{p}}} = \frac{\partial \sigma^{\mathrm{y}}}{\partial \bar{\epsilon}^{\mathrm{p}}} + \frac{\partial \sigma^{\mathrm{y}}}{\partial \dot{\bar{\epsilon}}^{\mathrm{p}}} \frac{\partial \dot{\bar{\epsilon}}^{\mathrm{p}}}{\partial \bar{\epsilon}^{\mathrm{p}}} = \frac{\partial \sigma^{\mathrm{y}}}{\partial \bar{\epsilon}^{\mathrm{p}}} + \frac{\partial \sigma^{\mathrm{y}}}{\partial \dot{\bar{\epsilon}}^{\mathrm{p}}} \frac{1}{\Delta t} = H' + \frac{\partial \sigma^{\mathrm{y}}}{\partial \dot{\bar{\epsilon}}^{\mathrm{p}}} \frac{1}{\Delta t} \qquad [6.119]$$

H' is the hardening modulus and the additional term exists in the case that the yield limit depends on the plastic strain rate.

Viscoplastic models require the solution of $X = \Delta t \cdot {}^{n+1}g\langle f \rangle - \Delta\bar{\epsilon}^{\mathrm{p}} \equiv 0$, Eq. [6.113]. The derivative with respect to effective plastic strain is

$$\frac{\mathrm{d}X}{\mathrm{d}\bar{\epsilon}^{\mathrm{p}}} = \Delta t \frac{{}^{n+1}\mathrm{d}g}{\mathrm{d}f} \frac{{}^{n+1}\mathrm{d}f}{\mathrm{d}\bar{\epsilon}^{\mathrm{p}}} - 1 = -\left[\Delta t \frac{{}^{n+1}\mathrm{d}g}{\mathrm{d}f}\left(3\,{}^{n+1}G + \frac{\mathrm{d}\sigma^{\mathrm{y}}}{\mathrm{d}\bar{\epsilon}^{\mathrm{p}}}\right) + 1\right] \qquad [6.120]$$

where the derivatives $\frac{\mathrm{d}\bar{\sigma}}{\mathrm{d}\bar{\epsilon}^{\mathrm{p}}}$ and $\frac{\mathrm{d}\sigma^{\mathrm{y}}}{\mathrm{d}\bar{\epsilon}^{\mathrm{p}}}$ in $\frac{{}^{n+1}\mathrm{d}f}{d\bar{\epsilon}^{\mathrm{p}}}$ are the same as in Eq. [6.118]. However, the viscoplastic models only use $\sigma^{\mathrm{y}}(\bar{\epsilon}^{\mathrm{p}})$.

Box 6.9 Evaluation of the derivative of the effective stress with respect to plastic strain increment

Eq. [6.111]

$$\Delta \boldsymbol{e}^{\mathrm{p}} = \Delta\bar{\epsilon}^{\mathrm{p}} \frac{3}{2\,{}^{\mathrm{trial}}\bar{\sigma}}\, {}^{\mathrm{trial}}\boldsymbol{s} \qquad [6.121]$$

and Eq. [6.110]

$${}^{n+1}\boldsymbol{s} = {}^{\mathrm{trial}}\boldsymbol{s} - 2\,{}^{n+1}G\Delta\boldsymbol{e}^{\mathrm{p}} \qquad [6.122]$$

are combined to

$${}^{n+1}\boldsymbol{s} = \left(1 - \frac{3\,{}^{n+1}G\Delta\bar{\epsilon}^{\mathrm{p}}}{{}^{\mathrm{trial}}\bar{\sigma}}\right) {}^{\mathrm{trial}}\boldsymbol{s} \qquad [6.123]$$

and thus

$${}^{n+1}\bar{\sigma} = {}^{\mathrm{trial}}\bar{\sigma}\left(1 - \frac{3\,{}^{n+1}G\Delta\bar{\epsilon}^{\mathrm{p}}}{{}^{\mathrm{trial}}\bar{\sigma}}\right) \qquad [6.124]$$

The derivative with respect to $\bar{\epsilon}^{\mathrm{p}}$ (or $\Delta\bar{\epsilon}^{\mathrm{p}}$) is

$$\frac{\mathrm{d}\,{}^{n+1}\bar{\sigma}}{\mathrm{d}\bar{\epsilon}^{\mathrm{p}}} = -3\,{}^{n+1}G \qquad [6.125]$$

7
Numerical methods and modelling for efficient simulations

The increase in the size of models (Fig. 1.1), is due to the continuous development of computer hardware and software. Some of the different numerical techniques and modelling choices that have been applied in CWM for increasing the computational efficiency are described below.

7.1 Element technologies

Considerable developments of efficient elements and solution algorithms for nonlinear heat flow and large deformations analysis have been undertaken. These general developments are not discussed here and the reader is referred to references given in Chapters 5 and 6. There is a development of element formulations performed specifically within the CWM community in order to alleviate the creation of a graded mesh. This is discussed in the next section.

One note about element formulation is in place here, concerning the consistency between the thermal and mechanical analysis [123]. The degree of the finite element shape functions for the displacements should be one order higher than for the thermal analysis. This is because the temperature field directly becomes the thermal strain in the mechanical analysis. The strains are obtained as the derivatives of the displacement field. Lindgren *et al.* [1] used eight-node, quadratic elements in the mechanical analysis and four-node, bilinear elements in the thermal analysis in a two-dimensional model of flame cutting. Usually the same elements are used in the thermal and mechanical analyses. The linear elements are preferred since smaller low-order elements perform better than larger high-order elements in nonlinear problems. However, they require reduced integration of the volumetric strain or enhanced strain fields [124] in order to avoid locking in case of large plastic strains due to the plastic incompressibility constraint. See Chapter 8 in [34] for a discussion of different elements and locking due to plastic incompressibility.

Friedman [125, 126] used a quadratic element but agreed with Hibbitt and Marcal [127] that a fine mesh with linear elements is to be preferred since linear quad, in two dimensions, and brick, in three dimensions, elements are the basic recommendation in plasticity [128, 129]. They perform better than linear triangles or tetrahedrons. Dhingra and Murphy [74] used quadratic (20-node) hexahedron elements in the modelling of a the bead on plate and tee flange welding cases. They used higher-order (10-node) tetrahedrons for a butt-welding case, with one layer of element over the thickness of the plate. Then the element is capable of representing pure bending as this has linear variation of the normal strain over the plate thickness. They obtained a good agreement with measurements for the thicker plates. If triangles or tetrahedrons are used in computational plasticity, then the elements should have quadratic interpolation functions for the displacement field. Andersson [88] used higher-order triangular elements. Using linear elements requires that the average temperature in each element should be used for computing a constant thermal strain in the mechanical analysis [123] for maintaining the consistency as discussed above.

7.2 Meshing

Perhaps the most difficult problem in CWM is that the scale of the relevant fields spans three or four orders of magnitude. The size of the upset zone, where plastic deformation has occurred, is typically in the order of centimetres. The weld deposit itself may be measured in millimetres. The HAZ, where phase transformations, stress and hydrogen can cause cracking, has structures varying over only fractions of a millimetre. Thus predictions of HAZ cracking may need length scales as small as 0.1 mm to capture details in the HAZ in a domain measured in lengths of 1 m to capture essential boundary conditions. One of the fundamental modelling guidelines in finite element methods is to use a fine mesh wherever needed and a coarse mesh elsewhere. This requires versatile mesh generators. Most mesh generators are based on schemes to generate triangle or tetrahedral elements. Then higher-order triangle/tetrahedral elements should be used in accordance with the discussion in the previous section. Some mesh generators post-process the tri/tet mesh into quad/hex mesh. There are also direct mesh generators for the latter case although they are not equally capable as the tri/tet mesh generators.

A review of mesh generation by Owen is accessible on the web [130] together with an overview of software [131]. The well-known conference, the Meshing Roundtable, focused on this theme, and other related conferences can be found on http://www.andrew.cmu.edu/user/sowen/mesh.html. Using mesh generators in dynamic or adaptive meshing implies that a new mesh is created. This mesh may be completely different from the previous one. Sometimes it may be sufficient to remesh only, i.e. taking an existing mesh and modifying it. The modification may be like splitting elements or reshaping elements. The thesis by Hyun [132] includes an evaluation of different remeshing and smoothing methods.

The mesh generation of hex elements is a challenging task when a graded mesh is needed. The development of the graded element by McDill *et al.* [124,

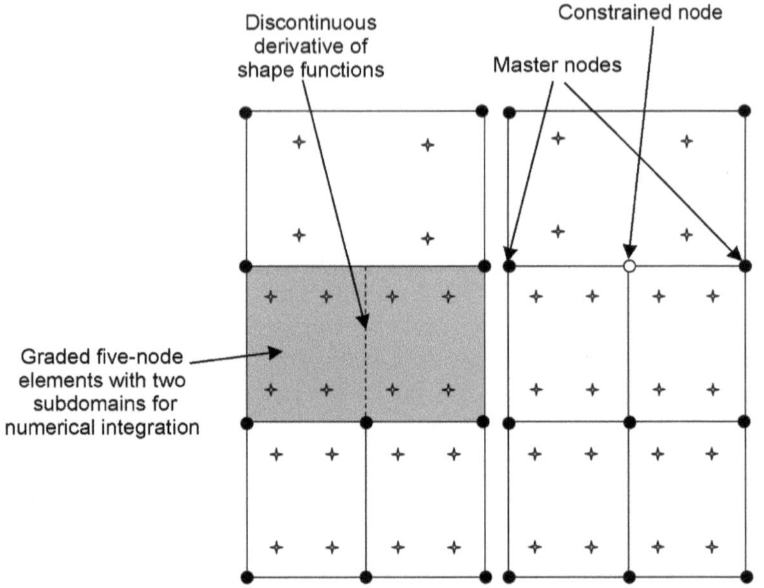

7.1 Graded mesh by graded elements or multiple point constraints. Nodes are denoted by circles and integration points by stars.

133] and graded shell elements [134] facilitates this. The elements have piecewise linear shape functions, enabling the creation of meshes like the one shown on the left in Fig. 7.1 with inter-element compatibility. The shape functions are piecewise linear and therefore the numerical integration is separated into domains with linear shape functions. The typically 2×2 Gausspoint rule is used within each domain, indicated by stars inside the elements. The same interpolation field with retained inter-element compatibility can be created by using constraints as shown on the right in Fig. 7.1. It should be noted that when plotting only the elements, it appears as though they have a larger domain with finer resolution. This is not the case, as the upper centre node is constrained to be on the straight line between its master nodes. (See also Box 7.1 for an illustration of the constraint method.) The graded shell element [134] is based on a solid-shell element formulation where only displacement degree of freedom is used, as later in [135]. This simplifies the connection with solid elements and

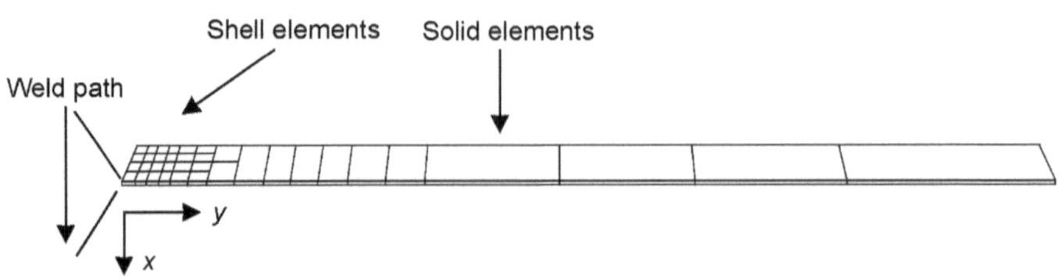

7.2 Solid-shell element from McDill *et al.* [134].

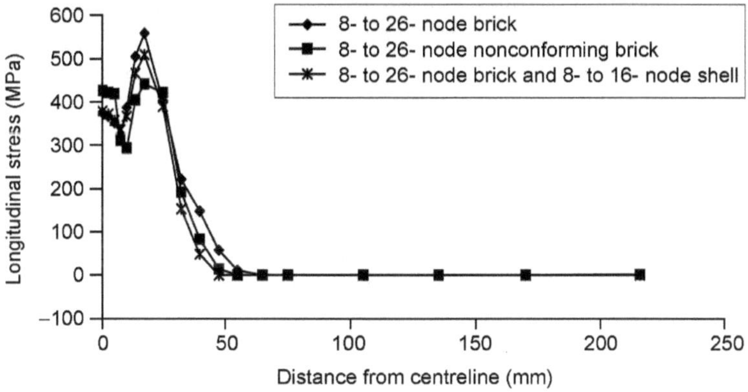

7.3 Residual longitudinal stress using solid-shell element from McDill *et al.* [134].

also has advantages in terms of plotting of results. A model using the solid-shell element near the weld is shown in Fig. 7.2 and corresponding computed residual, longitudinal stress is shown in Fig. 7.3. The latter figure also has computed results using two types of brick elements without any solid-shell elements.

Box 7.1 Formulation of constraint equation

The fulfilment of constraints of different kinds can be illustrated by means of the simple problem below. This is a general approach illustrated on a three-node one-dimensional element.

Assume that there is a heat conduction, steady state, problem to be solved using one three-node element. Assume furthermore that the centre node of this element should have the average temperature of the surrounding nodes. The total length of the heat conduction rod is L, the cross-sectional area is A and the heat conductivity is λ. Then the original 3-dof (degrees of freedom) problem can be reduced to a 2-dof problem as below. The original problem is

$$\frac{\lambda A}{6L}\begin{bmatrix} 14 & -16 & 2 \\ -16 & 32 & -16 \\ 2 & -16 & 14 \end{bmatrix}\begin{bmatrix} T_1 \\ T_2 \\ T_3 \end{bmatrix} = \begin{bmatrix} Q_1 \\ Q_2 \\ Q_3 \end{bmatrix} \qquad [7.1]$$

where Q_i is the nodal heat applied at the node number i.

The constraint equation is

$$\begin{bmatrix} T_1 \\ T_2 \\ T_3 \end{bmatrix} = \begin{bmatrix} 1 & 0 \\ 0.5 & 0.5 \\ 0 & 1 \end{bmatrix}\begin{bmatrix} T_1 \\ T_3 \end{bmatrix} = \boldsymbol{T}_{\text{constraint}}\begin{bmatrix} T_1 \\ T_3 \end{bmatrix} \qquad [7.2]$$

The constraint equation is applied to reduce the original problem to

$$T^{\mathrm{T}}_{\mathrm{constraint}} \frac{\lambda A}{6L} \begin{bmatrix} 14 & -16 & 2 \\ -16 & 32 & -16 \\ 2 & -16 & 14 \end{bmatrix} T_{\mathrm{constraint}} \begin{bmatrix} T_1 \\ T_3 \end{bmatrix} = T^{\mathrm{T}}_{\mathrm{constraint}} \begin{bmatrix} Q_1 \\ Q_2 \\ Q_3 \end{bmatrix} \qquad [7.3]$$

Carrying out the matrix multiplications gives the reduced problem

$$\frac{\lambda A}{L} \begin{bmatrix} 1 & -1 \\ -1 & 1 \end{bmatrix} \begin{bmatrix} T_1 \\ T_3 \end{bmatrix} = \begin{bmatrix} Q_1 + 0.5Q_2 \\ Q_3 + 0.5Q_2 \end{bmatrix} \qquad [7.4]$$

The centre node temperature is recovered after the solution of the above by the constraint

$$T_2 = \begin{bmatrix} 0.5 & 0.5 \end{bmatrix} \begin{bmatrix} T_1 \\ T_3 \end{bmatrix} \qquad [7.5]$$

The process above is also applicable for an incremental solution procedure and can be applied to any kind of constraint. The process consists of three steps:

- Set up constraint equations.
- Reduce matrices and vectors.
- Recover data for constrained nodes.

Sometimes the notation slave and master nodes is used in the above procedure. It can be applied to combine incompatible meshes, formulate substructures, etc. The process is usually described in terms of global equations but can be implemented on the element level. An example can be seen in CWM_Lab. A constraint matrix is set up to constrain the displacements of a node to average value of another node in the same element and a node in another element for enabling the creation of a graded mesh as in the right part of Fig. 7.1.

7.3 Dynamic and adaptive meshing

Dynamic and adaptive meshing are methods to concentrate the elements to regions with large gradients. Thus, increased computational efficiency can be obtained without sacrificing accuracy. Dynamic meshing denotes a technique where the user prescribed the refinement and coarsening of the finite element mesh. Typically a fine mesh region is moving with the heat source [136, 137]. Lindgren *et al.* [136] reduced the computer time by a factor of two using dynamic meshing. Lenz and Rick [138] obtained only a 20% reduction using dynamic meshing in the code Marc. Runnemalm and Hyun [139] decreased the computational time by a factor of two using adaptive meshing and Hemmer *et al.* [140] reduced the time by a factor of two to four.

The use of error measures to guide the mesh refinement and coarsening is a more accurate approach than dynamic meshing. Runnemalm and Hyun [139]

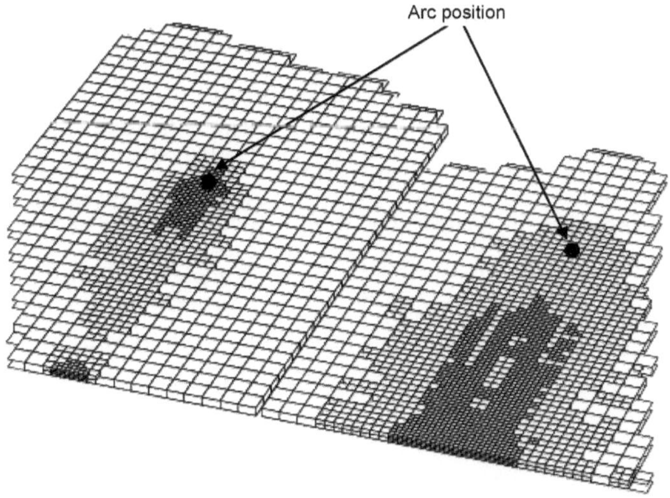

7.4 Zoom of mesh near arc when using temperature gradients, left, or gradient of effective stress, right, to control element sizes [139].

(see Fig. 7.4 for an example) found that it was advisable to combine thermal and mechanical measures when determining mesh refinement or coarsening and that the pattern for remeshing is not so easy to predict without an error measure. Rieger and Wriggers [141] discussed adaptive meshing using error estimators for coupled, problems where different meshes are used in thermal and mechanical analyses. More about error measures can be found in the book by Dow [142] other reference to publications about error measures can be found in [143]. The Zhu–Zienkiewizc's error estimator [144–147] is commonly used. The notation estimator denotes an error measure that can be related to error whereas error indicator is a kind of ad-hoc error measure. An example of the latter is to refine the mesh where the temperature is higher or the plastic strain is larger.

The different methods for remeshing are classified in Table 7.1 and the error measures they are recommended to combine with are given in Table 7.2. Combining the h-method with error indicators is a good approach.

Adaptive meshing can also be combined with distortion metric to avoid elements that are too distorted. This can be particular useful when simulating multipass welding [148]. Different error measures and adaptive strategies are discussed in Zienkiewicz *et al.* [147], Huerta *et al.* [149], Boroomand and Zienkiewicz [150] and Hyun and Lindgren [151] and the Oddy distortion metric [152], useful for remeshing when the mesh is distorted, is given in [152]. The metric is defined as:

$$D = \sum_{i=1}^{n} \sum_{j=1}^{n} C_{ij}^2 - \frac{1}{n}\left(\sum_{k=1}^{n} C_{kk}\right)^2 \qquad [7.6]$$

Table 7.1 Some types of remeshing techniques

	Description	Comment
Hierarchical h-method	Split elements	Easy to implement but cannot improve distorted elements.
General h-method	General remeshing	Complex to implement for two-dimensional quad elements and thee-dimensional hex elements. Software available for triangles and tetrahedrons.
p-method	Increase degree of shape function for element	Easy to implement. Induce a higher-order of continuity between elements. Must be combined with h-method if a lot of refinements need to be done.
r-method	Relocate existing node, keep element topology	Easy to implement but cannot create particularly graded mesh and has also limited capability to reduce distortion of elements.

Table 7.2 Appropriate combinations of remeshing methods and error measures

	Error indicator	Error estimator
r-adaptivity	✓⃝	⊗
h/p-adaptivity	⊖	✓⃝

where

$$C_{ij} = \frac{1}{\det|J|}\sum_{k=1}^{n} J_{ki}J_{kj} \qquad [7.7]$$

J is the jacobian of the mapping between the global coordinate system and the local coordinate system of the element. n is the dimension of the problem (2 or 3). The metric measure the deformation of the element (see Figs 7.5 and 7.6). The measure fails if J becomes negative. Then the elements have become inverted. A flowchart for remeshing accounting for element distortion and error measures is shown in Fig. 7.7.

7.4 Substructuring

Substructuring is an efficient tool for reducing the size of large models for linear, elastic analyses. Usually some approximations are introduced when implemented for nonlinear problems. The basic idea is to only include the region

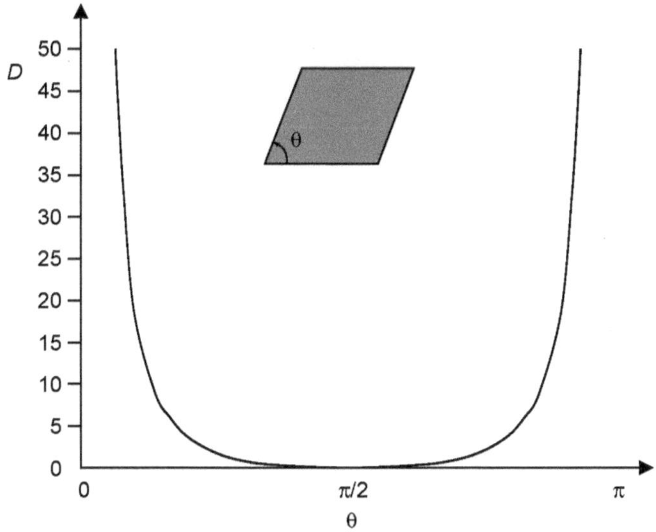

7.5 Oddy's distortion metric for shearing of element [152].

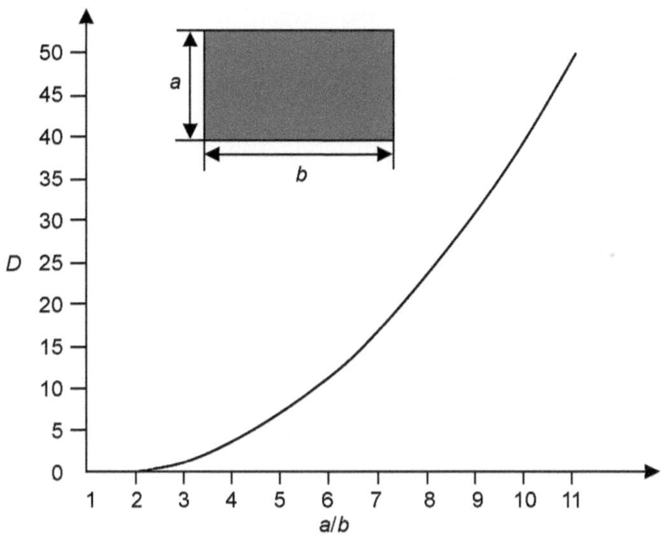

7.6 Oddy's distortion metric for elongation of element [152].

around the weld in the system of equations. The rest of the structure will be assumed to behave linearly and is condensed out. Thus one must ignore the temperature dependency of the material properties and the large deformations in the condensed part of the structure. This can be done by setting the boundary for this region where the temperature only will reach about 200–300 °C. This will not reduce the size of the problem so much as most elements are near the weld.

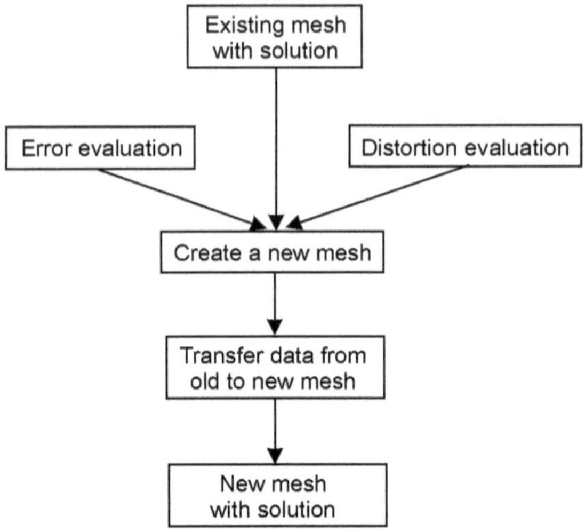

7.7 Flowchart for remeshing.

Therefore a dynamic substructuring is needed that is moving with the heat source. It is also necessary to obtain the thermal load from the condensed part to the analysed part. The thermal load cannot be obtained without doing a recovery step for obtaining deformations, stresses and thermal loads on this part and then some of the gain in speed is lost.

Andersen [153] combined substructuring and dynamic remeshing. A ship subassembly was modelled by linear, elastic shell elements. This global model was the basic information-carrying model (Fig. 7.8a). The panel had eight weld lines, as each stiffener was welded from both sides. A local solid model was created of the region whose welding was going to be simulated (see Fig. 7.8b). Dynamic remeshing was used for the local model. The stiffness of the global model must be increased between each weld pass, as there were initially only tack welds connecting web and plate. The tack welds corresponded to some shell element connections. The substructuring was not applied to the thermal model and there was no interaction between the thermal models for each weld, as it was assumed that the weld has cooled to room temperature before analysing the next weld. Andersen [153] was able to simulate a total of 8 m welding and studied the effect of the welding sequence on the residual distortion of the panel (see Fig. 7.9).

Souloumiac *et al.* [154] and Faure *et al.* [155] applied a similar local–global approach in combination with an inherent strain approach [156, 157]. They assumed that the residual plastic strains of the weld depend only on local mechanical and thermal conditions. However, they underlined the problem of knowing what boundary conditions should be applied to the local model as they influence the computed plastic strains. They also proposed a combination of local, transient macro-elements and steady-state (Eulerian) macro-elements to solve the local problem (Fig. 7.10). The nodal loads corresponding to the

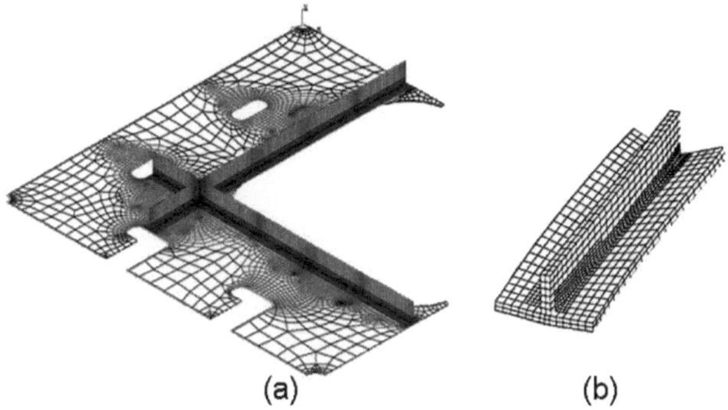

7.8 Global, shell model of ship subassembly (a) and solid model of part to be welded (b) [153].

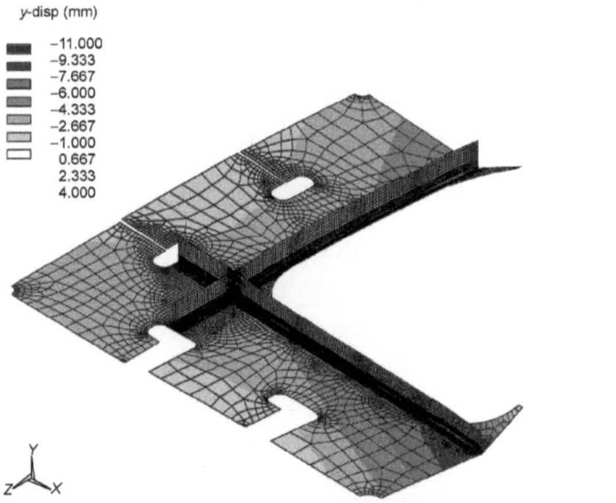

7.9 Deformed plate due to given welding sequence [153].

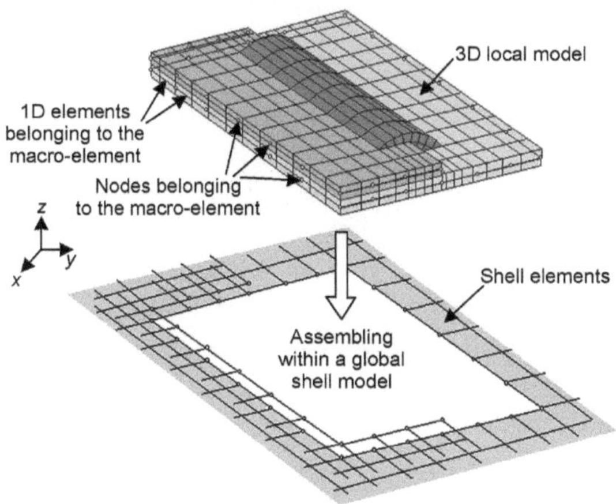

7.10 Macro-element definition [154].

computed plastic strains are then condensed to the boundary of the local model and there they are applied to the global model in order to compute the global behaviour.

7.5 Parallel computing

Parallel computing makes it possible to reduce the wall clock time for an analysis. Usually, implicit finite element methods are used in welding simulations and these are less scalable for parallel computing than explicit finite element methods. However, sometimes the limiting factor is computer memory and then it is possible to obtain a near-linear gain with the number of processors. Lenz and Rick [138] were able to reduce the wall clock time by more than a factor of two when going from one to two processors using Marc. Ericsson *et al.* [158][1] also analysed a large structure using Marc and parallel computing. Initial evaluations, not shown in their paper, using large models with solid elements gave a speed up to 1.7 times when going from four to eight processors.

7.6 Dimensional reduction

The most powerful strategy used to reduce the cost of thermomechanical simulations of welding has been to reduce the dimension of the problem from three to two or one. Typical weld models are shown in Fig. 7.11 and described in Table 7.3. The book by Radaj [7] contains detailed descriptions and reviews of the use of this kind of model. The residual stresses were predicted as if by the collection of many parallel uniaxial specimens. The earliest two-dimensional finite element analyses appeared in the early 1970s, by Iwaki and Masubuchi [159], Ueda and Yamakawa [160, 161], Fujita and coworkers [162, 163] and Hibbitt and Marcal [127]. Lindgren [86] outlines the development of the models from these first models up to 2000. The first three-dimensional residual stress predictions of full welds appear to be Lindgren and Karlsson [164], Karlsson and Josefson [75] and McDill *et al.* [165].

There is a hierarchy of elements (see Section 9.13) where each choice is imposing a constraint on the solution. The most general element is the solid element. The shell elements can be applicable for thin-walled structures provided detail near the weld is not needed. A combination of solid and shell elements is an alternative. All other choices reduce the deformations to a subset of three dimensions. Goldak *et al.* [166] discussed the kinematic constraints both for the thermal and the mechanical models when introducing two-dimensional assumptions. Andersson [88] showed that neglecting the heat flow in the welding direction is not a large error in the thermal analysis. The kinematic constraint introduced by assuming plane strain gives larger residual, longitudinal

1. Private communication.

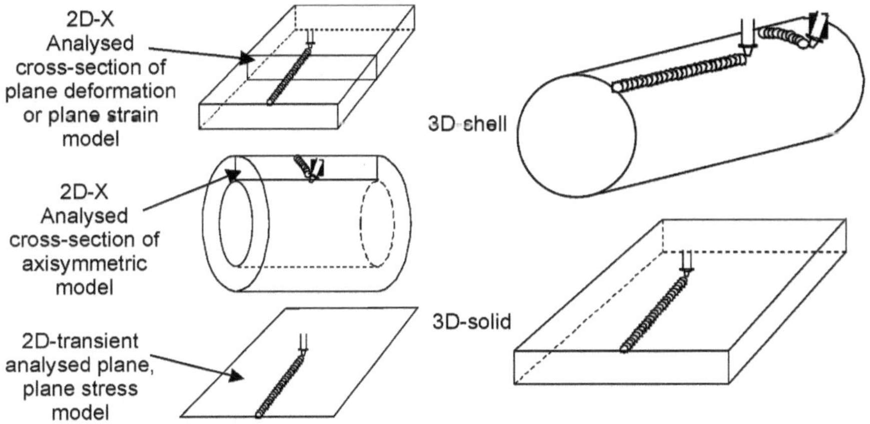

7.11 Two- and three-dimensional models of welding.

Table 7.3 Description of different geometric models

1. 2D-X, ignores the heat conduction in the welding direction. The model corresponds to plane strain, generalised plane deformation and axisymmetric models.
2. 2D-P[a], assumes constant temperature over the thickness. The model is a plane stress model where the heat source is moving in the plane of the model.
3. 3D-shell[a] can have varying assumptions about possible temperature variation over the thickness are possible. The total bending strain varies linearly over the thickness according to shell theory.
4. 3D-solid, the most general case.

[a] Multipass welds can be accommodated approximately by changing thickness of the elements.

stresses, e.g. [167]. Assuming plane deformations can alleviate this, which can work well for slender structures where the stresses in the weld make them behave like a beam [168].

The cross-sectional models, denoted 2D-X, correspond to assuming an infinite welding speed when simulating a circumferential butt weld on a pipe using the axisymmetric model. The same result would be obtained with a 3D-solid model of the pipe using and applying the weld around the whole circumference simultaneously. It is also possible to consider the plane strain model as a weld with infinite speed. However, it will not correspond to using a 3D-solid element model with the weld laid instantaneously. This is due to the kinematic zero normal strain constraint in the mechanical analysis. This condition means that the net longitudinal residual stress after cooling is not zero. It is as if the whole plate was rigidly fixed in the longitudinal direction. The too high longitudinal restraint due to the plane strain assumption can be alleviated by assuming generalised plane strain [88], or plane deformation [169–175]. Michaleris *et al.* [176] compared two- and three-dimensional models of a multipass butt-welded plate. They simulated the first pass with both models. The two-dimensional model had a larger fusion zone, despite a reduced heat input, and a larger zone with high tensile longitudinal stresses than the three-dimensional model. Ferro *et al.* [177] compared a plane deformation model (2D-X) and a 3D-solid model and obtained a good agreement between their residual stress fields.

Runnemalm and Berglund [178] compared the use of 2D-X, 3D-shell and 3D-solid models of a butt-welded plate (see Table 7.3 for these notations). The plate is shown in Fig. 7.12 discretised by 3D-solid elements and by shell elements in Fig. 7.13. The 2D-X model with plane strain assumption is shown in Fig. 7.14. They performed different comparisons of the results at the locations numbered in the figures below and along the reference line.

Berglund and Runnemalm [178] applied three different sets of boundary conditions. The deviation of the deformation result between the 3D-solid model and the other two models is smallest in the case with fixed edges parallel to the weld line. This is the case with largest restraint in the model. It is not possible to predict the transient behaviour in the 2D model because it corresponds to infinite welding speed in the thermal analysis, although accurate prediction of temperature history perpendicular to the welding direction is possible. Neither can the model capture some of the deformation modes. One example of comparisons in their paper is the transverse displacement at point 1 shown in Fig. 7.15. The difference they found between the 3D-shell and 3D-solid elements may be reduced provided the thermal loading had been calibrated in the 3D-shell model. They found, like others, that the longitudinal residual stress is higher in the 2D-X model than in the other models.

Dye *et al.* [179] also compared different geometric models for a plate problem. They also compared elements with quadratic and linear shape func-

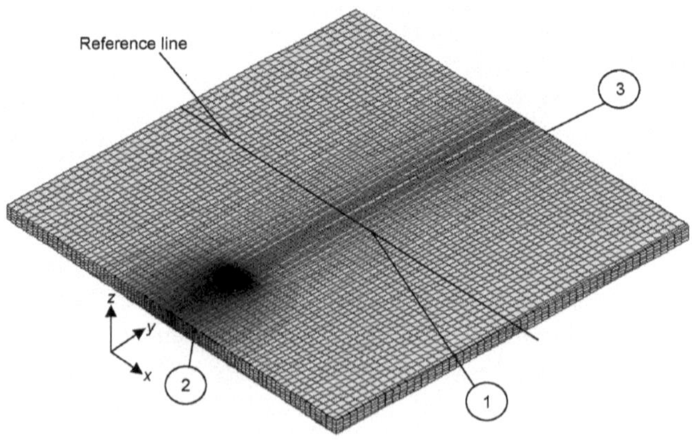

7.12 3D-solid model [178].

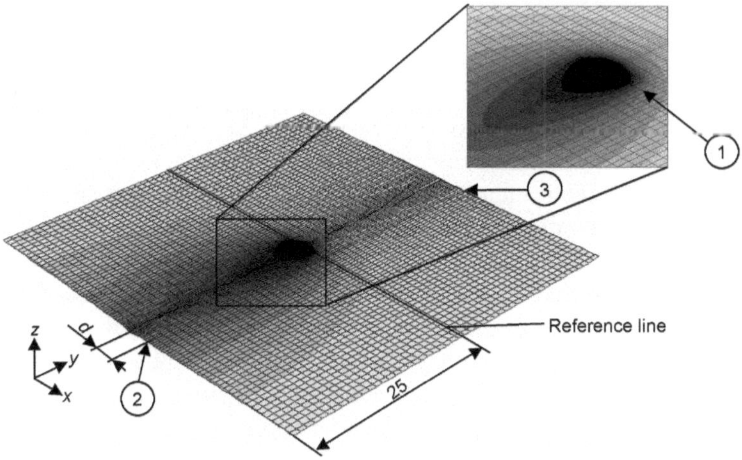

7.13 3D-shell model [178].

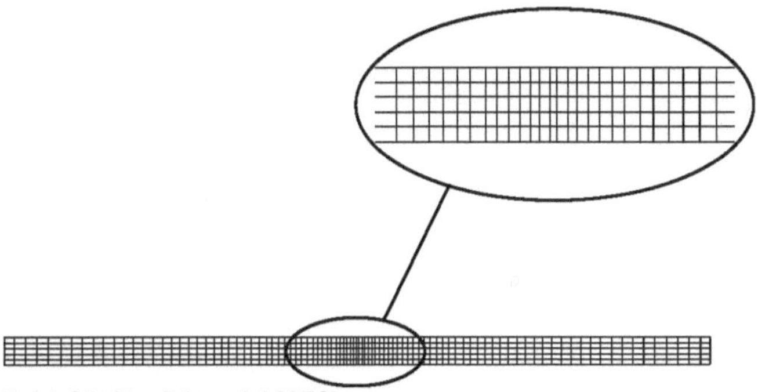

7.14 2D-X solid model [178].

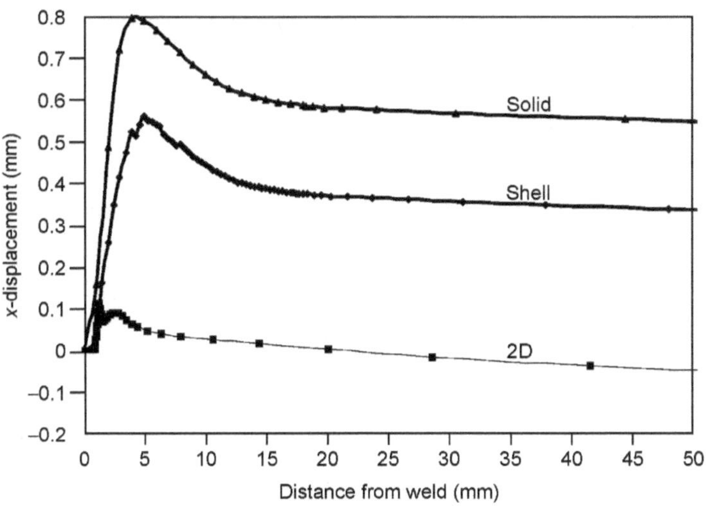

7.15 Transverse displacement at point 1 in the different models for the case of fixed edges parallel to weld line [178].

tions. They chose an element with linear shape functions and additional incompatible bending modes for the 3D-solid elements.

Many investigations of circumferential welds on pipe have been performed. Ravichandran *et al.* [180] reviewed some papers. Most work before the mid-1990s was based on using axisymmetric models, but since then there have been an increasing amount of publications with three-dimensional models. The axisymmetric models cannot represent the transient response during the welding as they correspond to an infinite welding speed. They have been used to obtain quite good residual stress fields. The three-dimensional models are necessary to obtain the correct circumferential distribution and the start–stop effect of the residual stresses.

The axisymmetric models have been useful for obtaining good residual stress fields but they are not able to represent the deformations correctly. For example, the axial shrinkage will very likely be wrong when using an axisymmetric model because it corresponds to having the whole circumference melted at the same time. This will affect the computed axial shrinkage. Zhang *et al.* [181] used axisymmetric, 2D-X, model for computing welding residual stresses due to a multipass weld. They mapped this field to 3D-shell model of a quarter of the circumference of the pipe. This model was used to study the effect on the stresses by removal of material. Dong *et al.* [182] compared residual stresses from a 3D-shell model with measured values. They obtained a quite good agreement. Fricke *et al.* [183–185] used a three-dimensional solid model and although the model had a very coarse mesh the analysis was very time consuming. Dike *et al.* [186] showed the circumferential variation of the residual stresses. The variation and the start–stop effect can be clearly seen in Fig. 7.16. They compared with measured values at 135 °C from the weld start and along the axial direction. They obtained good agreement but did not check the circumferential variation with measurements.

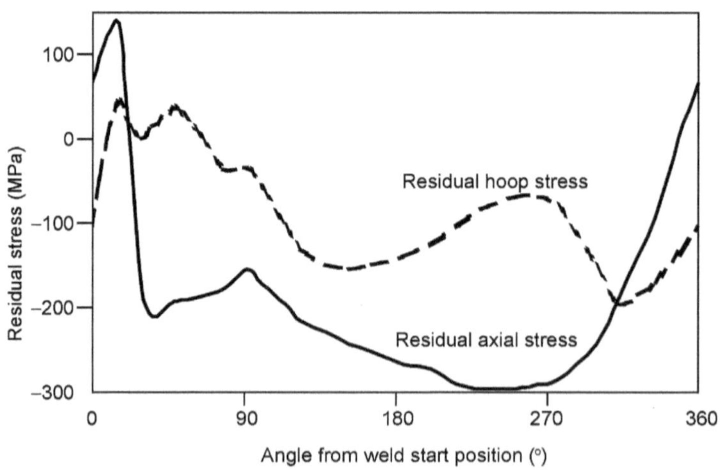

7.16 Distribution of residual stresses for circumferential weld in a pipe [186].

The cases of deformation due to several one pass welds in panels [176, 187] and panels with stiffeners [153, 188, 189] have received some attention. Chakravarti *et al.* [187] reduced the geometry into a thin slice of the structure. This made it impossible to include all deformation modes into the model. The comparisons that could be made between measured and computed residual deformations and stresses were good. Michaleris *et al.* [176] tried to overcome the need for a three-dimensional welding simulation by extracting load of weld from a two-dimensional model and applied it to a three-dimensional model (see Section 7.8). Dong *et al.* [190] proposed the use of 3D-shell elements with different integration layers, as in the case of modelling composites, for modelling multipass welds.

7.7 Weld pass reduction

The reduction of dimension corresponds in the 2D-X models to assuming infinite welding speed (Section 9.13). It is also possible to lay a weld instantaneously, i.e. assume infinite welding speed, without reducing the dimension of the geometry. This is one approach to reduce the effort to simulate a weld. It is also possible to merge or lump several weld passes into one weld pass. This corresponds to simulating a case with large heat input and more filler material in fewer strings than in the original problem.

Rybicki and coworkers [191–193] studied this and applied a very coarse simplification of the temperature field using an envelope method. The envelope method extracted an envelope of temperatures from the thermal analysis to be applied in the subsequent mechanical analysis; see their papers for details. Many papers about multipass welding followed this initial work and there were also papers focusing on the simplification aspects for obtaining residual stresses in multipass welding [194–196]. Lindgren [86] reviews these approaches. However, it is quite possible to perform a complete analysis of large multipass welds using two-dimensional models [3] without using lumping or envelope techniques. The discussion about weld pass reduction is of interest when using three-dimensional models. The use of three-dimensional models is still very limited for multipass welds [183, 184, 197–200]. Dike *et al.* [200] found that a two-dimensional model (Fig. 7.17) could not predict the transverse shrinkage of a steel plate with a four-pass weld (Fig. 7.18). They had to use a three-dimensional solid model. They also found that it was possible to lump the welds and obtain a quite good residual deformation. However, the problem of infinite welding speed that is implicitly assumed in the two-dimensional model also shows up in the three-dimensional model if the whole weld is laid instantaneously. This indicates that extra care is needed when choosing the geometric model if the deformations are of interest. This is probably due to the effect of the changing weld restraint during the welding.

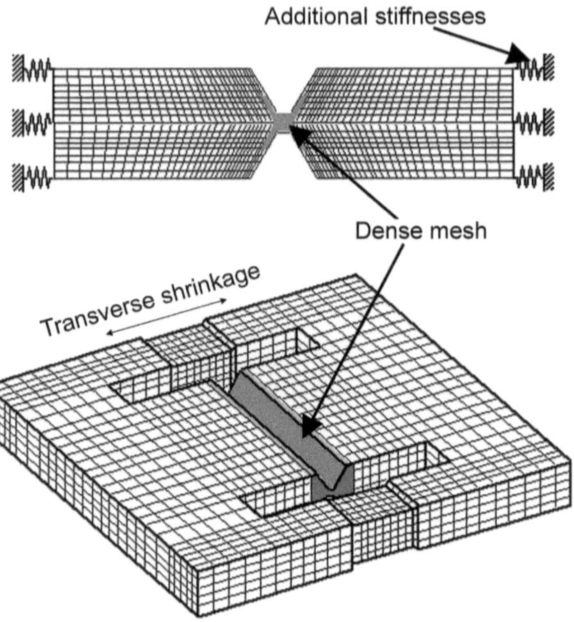

7.17 Two- and three-dimensional models of experimental set-up for measuring transverse shrinkage [200].

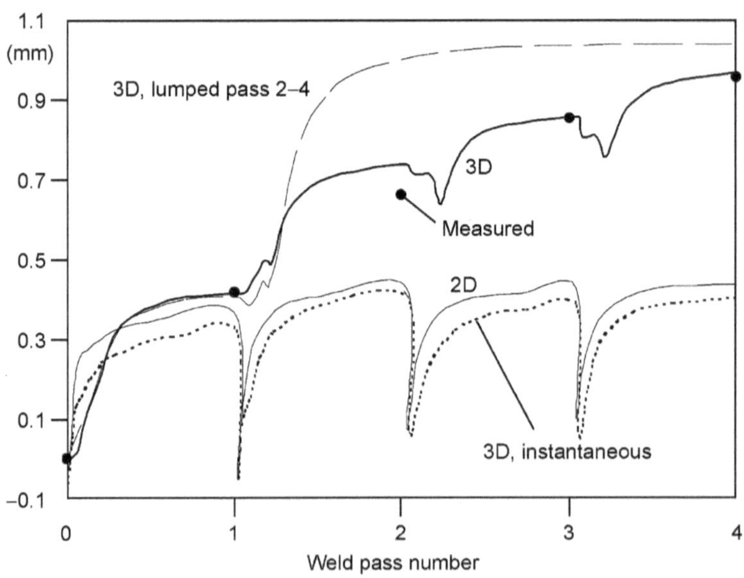

7.18 Computed and measured transverse shrinkage [200].

7.8 Replacement of weld by simplified loads

Another technique for reducing the computational effort required in the simulation of one pass welding, or more often multipass welding, involves translating and superimposing the residual stress or inelastic strain field from a single pass to the other passes. It is not possible to directly superimpose the stresses. The stresses have corresponding nodal forces in the finite element models. These forces can be superimposed in a separate analysis that gives the wanted (approximate) total stress field.

Overlay weld repair consisting of several hundreds of weld passes was studied by Chakravarti *et al.* [187, 201]. The residual stress pattern of the multipass weld was created by superimposing the residual strains of individual welds and gradually building the welds to patches of welds and finally combining all these patches. The total distortion is determined as the elastic response of the structure to this accumulated stress field.

Michaleris and DeBiccari [189] used a two-dimensional model for a detailed welding simulation (upper part in Fig. 7.19). They transferred welding residual stresses obtained from this plane deformation model to a three-dimensional

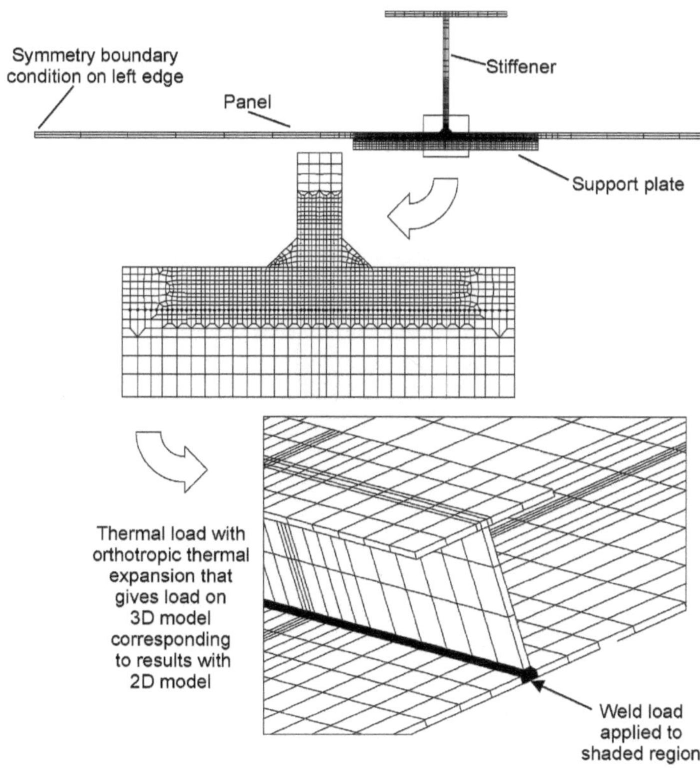

7.19 Models used in Michaleris and DeBiccari [189].

model (lower part in Fig. 7.19) by means of thermal strains. A constant temperature in the weld region was imposed on the latter model together with use of orthotropic thermal expansion. This model of the structure was then used to study weld-induced buckling. Thus it was possible to achieve different loading in the longitudinal and transverse directions of the weld. The thermal load was chosen so that the stresses in the midspan of the three-dimensional model became the same as in the two-dimensional model. This approach has been used by others [188, 202, 203] and may be called an inherent shrinkage method.

The inherent strain method developed by Ueda and coworkers [156, 157, 204] has roots back to work in Germany by Schimmöller [205, 206]. Radaj [7] give a very good formulation of the method and its history. The basic idea is simple. Performing a complete CWM simulation leads to the residual stress state where the stresses are computed from the elastic strains. The total strain field is the sum of elastic and inelastic strain fields. There is a compatibility requirement on this field [28, 97]. This cannot, in the general case, be fulfilled by the inelastic strains alone. Therefore, the presence of inelastic strains causes elastic strains and the latter give rise to residual stresses. Thus, knowing the residual inelastic strains makes it possible to compute residual stresses. These inelastic strains are named inherent strain by Ueda and others. The inherent strain method shares the same idea as the approach described above where the thermal strains play the role of the inherent strains. The inherent strains can be obtained from measurements, simple models or CWM simulations. However, the latter do not give any simplification in the procedure if it has to be done for every case. Assuming that the residual inelastic strain only depends on local conditions would make it possible to set up a set of inherent strain fields corresponding to a specific welding procedure. It could then be applied to different structures.

Yuan and Ueda [207] investigated the inherent strain for welded T- and I-joints. The problem of determining the relation between the inherent strain and the welding parameters is addressed by Wang *et al.* [36]. Murakawa *et al.* [157] described the method and showed different applications. Jung and Tsai [208] also mapped residual plastic strains into an elastic model for obtaining deformations. They denote the approach plasticity-based distortion analysis (PDA).

8 Calibration and validation strategy

The assessment of the correctness and accuracy of simulation results is extremely important in any practical application. The awareness of this is obvious in CWM as can be seen by the large number of papers including both experimental and computational results. Lindgren [209] lists a large number of publications up to 2001 with this content.

The aim of this chapter is to propose a strategy to perform this assessment in a structured manner. The validation issue is closely related to the calibration of the model, which is also necessary in CWM. It is important that validation and calibration are kept distinct otherwise a model may be considered validated although it is only calibrated.

The concepts introduced above and other relevant notations are first defined. The aim is to draft a combined calibration and validation strategy useful within the scope of classical CWM models. The chapter is closely related to Chapter 9 where a methodology for creating models for different scopes within CWM is described.

8.1 Definitions of concepts used

A *model*[1] is a theoretical or experimental device conceived to describe certain aspects of phenomena, processes and/or objects in the real world. A theoretical model consists of symbolic information, i.e. executable data and procedures. Its purpose is to predict real world behaviour under defined real or hypothetical conditions. Real world behaviour is generally extremely complex; model behaviour (also termed 'virtual world behaviour') is greatly simplified. A distinction can be made between the conceptual model and the mathematical models. The *conceptual model* is the idealisation or simplification of the real world system that must be done as a first step in the modelling process. The conceptual model is sometimes called a physical model as one important

1. http://www.hyperdictionary.com/search.aspx?Dict=&define=model

ingredient in its creation is the determination of which physical phenomena are important to include in the model. This conceptual model is then expressed in a *mathematical model*. Several mathematical models are embedded in a finite element code. The creation of a *finite element model* defines the set of mathematical models that will be executed during a simulation. The different choices of element types, material models, boundary conditions, etc. correspond to different mathematical models. Finally, the solution is obtained from the finite element model. Figure 8.1 illustrates this and some of the other definitions below. The role of the conceptual model has been omitted: it is the mathematical model that is used as a basis for the qualification process.

Those aspects of a real world phenomenon that are in focus when evaluating a model are the *scope*[2] of the model. Defining the scope of a model is the most crucial aspect in CWM. It is mandatory to know why a welding process is to be simulated. The scope and at what stage in a development process the model is to be used sets the requirement of accuracy of the model as discussed in Chapter 10.

The term *simulation*[3] is often used synonymously with modelling, but there are differences in meaning: 'A simulation should imitate the internal processes and not merely the result of the thing being simulated'. This gives an association into a simulation as a model that imitates the evolution in time of a studied process. For example, a simplified model directly giving residual stresses due to a welding procedure will not qualify as a welding simulation. However, the term 'simulation' is often, and also in this book, used to denote the actual computation. Simulation errors will then be those errors related to the solution of nonlinear equations as well as the time stepping procedure.

The concept of verification and validation (V&V) defined below is taken from the field of computational fluid dynamics. The papers by Oberkampf and coworkers [210–213] provide the basis for the concepts used in the following. The term *verification* is the 'substantiation that a computerized model represents a conceptual model within specified limits of accuracy' [210]. The conceptual model is expressed by a mathematical model in the current context. Thus verification is defined as the process of confirming that the equations defining the mathematical model are solved correctly and with expected convergence properties. It requires stable and consistent numerical algorithms [34, 85, 214] and is the checking that the solution of the finite element model is correct with respect to the general mathematical model embodied in the code. This is the code verification process when different analytical solutions, benchmark cases and other numerical tools are used to check the computer code. The user of commercial software often assumes that this has been done already by the software developer. However, the complexity of software development and the

2. http://www.hyperdictionary.com/search.aspx?define=scope
3. http://www.hyperdictionary.com/dictionary/simulation

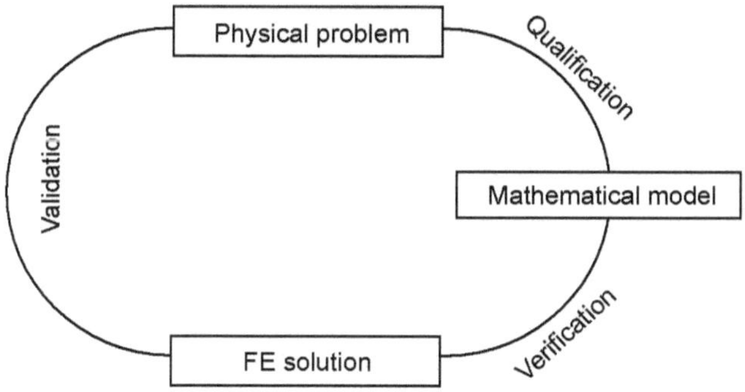

8.1 Validation and verification in finite element modelling [14].

large number of options that can be combined in the modelling process are reminders that this cannot be taken for granted. Verification is not discussed. Some examples of verifications are integrated with the finite element code, CWM_Lab, available to purchasers of the book (see pages xiii–xiv).

Qualification of the model is the determination of adequacy of the conceptual model to provide an acceptable level of agreement for the domain of intended application. The conceptual model is replaced by the mathematical model in Fig. 8.1 as the latter is usually the expression of the first model. The *modelling errors* are evaluated in this process. This is discussed in Section 8.3.

Validation is the checking of the accuracy of the model with respect to real world behaviour. Thus if verification is the checking that the equations are solved correctly, then validation can be considered as the checking that the correct equations are solved. The validation process is strongly related to the scope of the model and the accuracy needed. The aim is to create a sufficiently valid and accurate model. The validation is done by comparison with experimental results from the studied problem. A validation metric that also accounts for the uncertainty in the measurements is preferable [210]. The *simulation errors* are evaluated in this step. It also encompasses evaluation of the discretisation errors. Simulation errors also include solution errors in the solution process of the nonlinear system of equations as well as parametric errors. The latter concern the accuracy of the numerical values used for model parameters. A quantitative evaluation of errors is preferred to a subjective estimate. Sudnik *et al.* [215] quantified the parametrical errors, uncertainties, in the material parameters used in a weld process model and their effect on the predicted weld pool dimensions. Alberg [216] use a validation metric from [210] without including the effect of scatter in measurements.

Belytschko and Mish [217] discuss different types of uncertainties and the problem of creating a valid model. Errors are introduced from uncertain input data due to insufficient measurements. They call these epistemic or reducible errors [213]. Other errors are irreducible errors due to stochastic variations. This is also related to the question about robustness, which is discussed in Section

11.1. The concept of *computability* is introduced by Belytschko and Mish as an estimate of how well a process can be modelled.

The use of measurements to determine necessary input to model parameters, such as material properties, is called *calibration*. Specific parts of the model, submodels, are calibrated in order to fit the measurements. Thus, a calibrated model will fit the experimental data used in this process. It is important to maintain the distinction between calibration and validation [218]. Therefore, validation of the model must be done on other experimental results than those used in the calibration process.

Prediction is the modelling and simulation of a specific case that is different from the validated case. Oberkampf *et al.* [211] discuss the problem of 'how nearby' the prediction case is to the validated case. This determines how confidently the accuracy of the results from the predicted case can be deduced from the validation case, and depends on how large the overlap between the two cases is. If the same phenomena dominate the behaviour of interest in the prediction case as in the calibration and validation steps then the confidence in the predicted results is high.

Figure 8.2 shows more in detail the relations between the different concepts discussed above. This figure will be explained in more detail in Section 8.5 where a combined calibration and validation is proposed.

8.2 Code verification

Adding a given mathematical model to a finite element code can be quite a straightforward task: for example, if it is a new constitutive model or a new logic

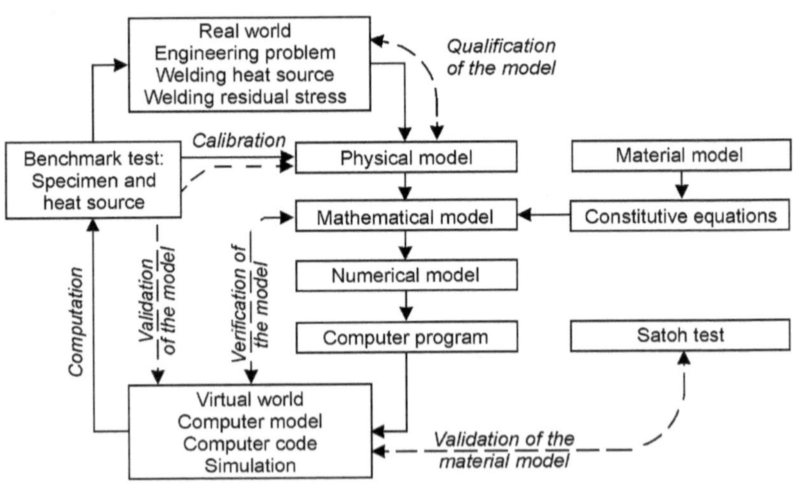

8.2 Flowchart of calibration, verification, validation and modelling in CWM [219].

for contact searches. The verification of this particular, isolated, piece of code is a larger piece of work including checking its stability for a variety of input data. However, the verification of the complete software is a gargantuan task. Unexpected interactions and combinations can cause problems, despite a careful modular approach in the software development. This problem is not addressed here. There are a variety of tests to control the consistency and stability of the implemented algorithm. The patch test, rigid body deformation test and benchmark cases are tests that can be implemented in the software for automatic checking the performance of the code which has to be repeated as soon as new features are added. The CWM-Lab software has only a limited set of automatic tests implemented. The NAFEM organisation (http://www.nafems.org/) has proposed different tests for ensuring the quality of software and also the quality of its users.

8.3 Model refinement and qualification

The most important step in creating the conceptual, physical, model is to understand the physics of the problem at hand (Chapters 2–4) and to know why a simulation is considered. Only then is it possible to judge what phenomena have to be included in the model – the scope of the model. However, sometimes it can be appropriate to start with a simpler model than the scope requires in order to gain familiarity with the problem at hand. The physical modelling is the step of defining by what simplified physical processes the real world behaviour will be substituted, e.g. continuum mechanics and thermodynamics, heat conduction, thermoelasticplastic material behaviour and heat source definition. These are always, in the CWM context, expressed by means of a finite element model. This is the focus of Chapter 9. Therefore, the development of the finite element model should start with a clear picture of the scope of the model. The model descriptions in Chapter 9 are based on the assumption that the model is within the framework where classical CWM is applicable.

The model qualification – comparison of the conceptual model with the real engineering problem (upper dashed line in Fig. 8.2) – is the first step in the modelling process as stated in the previous paragraph. This is an abstract judgement of the sufficiency of a possible conceptual model to capture the relevant phenomena. Despite careful initial considerations it may turn out at the validation stage that the originally chosen conceptual model is not sufficient. Then a more general physical model is chosen. This requires access to a sequence of models. Each new level of conceptual model includes the physics of the previous level plus the inclusion of some additional phenomenon. Some examples of this are given in Table 8.1 from Radaj and Lindgren [219].

Table 8.2 lists constitutive models of varying complexity that may be considered in CWM simulations. This is just one example as there is no strict hierarchy in which features are added to the constitutive model. For example, it

Table 8.1 Example of two levels of model refinement from Radaj and Lindgren [219], for refinement options for thermal (th) and thermomechanical (tm) models of seam welding

Simplified model	Refined model
2D cross-sectional model (th, tm)	3D solid element model
Momentary area heat source (th, tm)	Moving volume heat source
Flush weld geometry (th, tm)	Reinforced weld geometry (overfill)
Uniform conductivity (th)	Increased conductivity in weld pool
Uniform heat capacity (th)	Modified heat capacity (phase change)
Integrated filler material (th, tm)	Dummy elements for filler material
Uniform elastic-plastic properties (tm)	Non-uniform elastic-plastic properties
Without transformation behaviour (tm)	With transformation behaviour
Temperature-independent material parameters (th, tm)	Temperature-dependent material parameters
Restriction to cooling process (tm)	Heating and cooling process
Adiabatic boundary conditions (th)	More realistic thermal boundary conditions
Rigidly restrained boundaries (tm)	More realistic boundary conditions
Lumped layer approach (th, tm)	Multilayer approach
Without weld start and end (th, tm)	With weld start and end

is possible to use a model with a combined isotropic and kinematic hardening model without inclusion of phase changes. Then this model has no natural level in the structure of Table 8.2 where it can be inserted.

Another example of conceptual models of increasing capacity is expressed in the choice of different finite element types. Figure 8.3 shows their hierarchy. At the top is the most general model where no simplifying assumptions about the deformations have been made. Stepping down the diagram reduces the dimensions of the problem. The so-called continuum mechanics elements are on the left side and so-called structural finite elements are on the right. They assume a linear variation in the in-plane strain over the thickness of the element. The element types have corresponding mathematical formulations for the heat flow. These are not discussed here but are given in any textbook about finite elements [32, 34, 83] or respective continuum mechanics [28, 97]. The choice of elements for different welding situations and the physical implications of these choices are given in Section 9.13.

8.4 General approach for validation

Verification and validation (V&V) is the primary method to assess accuracy and reliability in computational simulations. The foundation for this strategy presented here is the papers by Oberkampf *et al.* [210–213]. The methodology is based on the notion that the complete system is too complex and/or expensive to use for validation tests. Therefore, the validation strategy employs a hierarchical methodology that segregates and simplifies the physical and coupling

Table 8.2 Constitutive models of varying complexity

Name of model	Details
Rate-independent plasticity material model	Temperature-dependent coefficients Isotropic, linear hardening
Rate-independent plasticity material model with predefined phase changes with their effects on thermal dilatation and yield stress	Above plus different set of curves for thermal dilatation and flow stress during heating and cooling and a criterion for combining them
Rate-independent plasticity material model with predefined phase changes with their effects on thermal dilatation and yield stress with TRIP	Above plus transformation induced plasticity in case of martensite/bainite formation
Rate-independent plasticity material model with microstructure model and mixture rules	Above plus evolution equations for microstructure
Rate-dependent plasticity material model with microstructure model and mixture rules	Above plus rate-dependent plasticity
Rate-dependent plasticity, advanced hardening, material model with microstructure model and mixture rules	Above plus combined isotropic and kinematic hardening

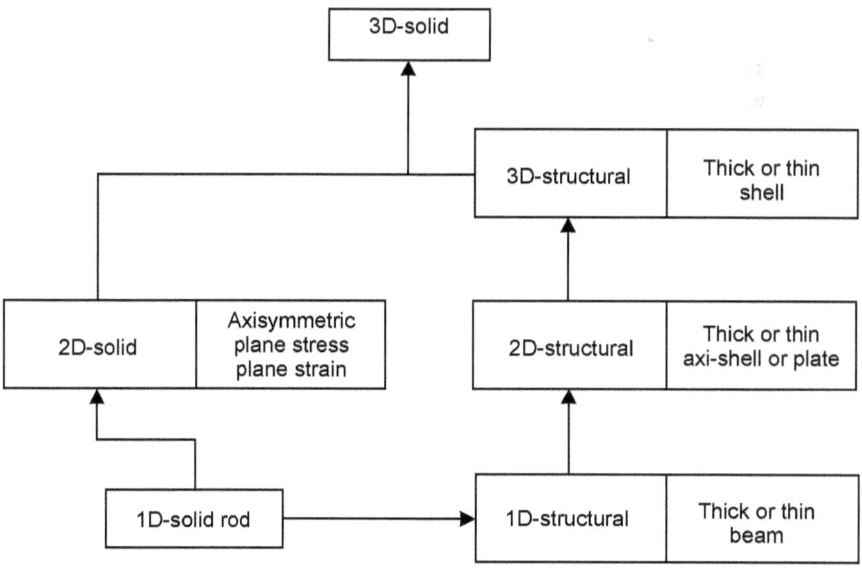

8.3 Hierarchy of element formulations in finite element modelling.

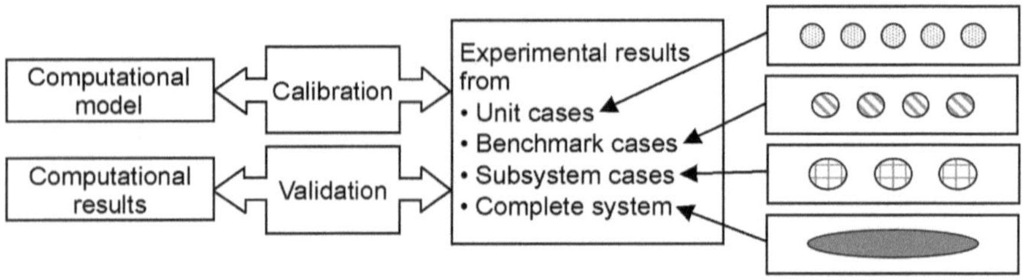

8.4 Verification and validation process (left) and structure (right).

phenomena involved in complex engineering systems. The hierarchy is decomposed into four levels, system level, subsystem level, benchmark level and unit problem level (right side of Fig. 8.4). This decomposition is used to isolate particular phenomena. The proposed strategy for validation in CWM cannot be strictly hierarchical owing to the couplings that exist [219]. One example is that it is not sufficient to redo the validation of the material model when changing material. It also becomes necessary to repeat the validation of the heat source model as well even if the same welding procedure is used. The concepts of unit problems and benchmark cases play the major role in the proposed calibration and validation strategy in CWM.

The complete system consists of the actual engineering hardware for which a reliable computational tool is needed. All the physical and geometrical effects occur simultaneously in this system. Measured data refer to the engineering hardware under realistic operating conditions. The quantity and quality of such measurements are costly and naturally very limited. Subsystem cases represent the first decomposition of the actual hardware into simplified systems or components. The complexity of physics and geometry is reduced. The quantity and quality of measurements are enhanced, but are still limited. It should be noted that many applications in CWM start at the subsystem level. Typically the welding of a component of a larger system is analysed. However, it is possible to have, for example, a ship structure with a large amount of plates, stiffeners, etc. that are welded. This can then be considered the complete system and a specific set of welds can be a subsystem.

Benchmark cases represent the next level of decomposition: special hardware is fabricated to represent the main features of each subsystem. Physics and geometry are substantially simplified with focus on those parameters that have great influence on the results. The quantity and quality of measurements should be very high.

Unit problems represent the baseline decomposition of the complete system. Only one element of complex physics is allowed to occur in each of the unit problems that are examined. Unit problems are characterised by very simple geometries. These types of measurements are commonly conducted in specialised research laboratories.

8.5 Calibration and validation strategy

Material modelling is always a problem when modelling nonlinear deformations [217]. This is particularly true in the case of thermomechanical problems where the material behaviour (see Section 9.7 for details) involves a large range of complex physics. The other problem, which is specific for CWM, is the heat input model. These models need to be calibrated as well as validated. The focus regarding the heat input model is, in the following, on the distribution of heat, although it should be noted that the modelling of the weld region also has other simplifications. The basic structure of the proposed calibration and validation is shown in Fig. 8.5. Some of the tests described in the following are defined as benchmark cases although they have a simple geometry and therefore could be considered as unit cases equally well. The motivation for the current choice is that they exercise a range of phenomena.

The unit cases (upper right box in Fig. 8.5) are purely used for calibration purposes. The benchmark cases (middle box to the right in Fig. 8.5) are used to calibrate the heat input and for general validation of the complete model. If the validation also includes a subsystem case then this is for validation purposes only. It is assumed in the proposal that it is never necessary to resort to validation on the complete system level.

8.5.1 Calibration and validation of material model

The material properties are often found in the literature, or must be obtained from experiments. The latter issue is not discussed below. The requirements for material models are related to the different accuracy levels given in Chapter 9. Thus the range of validity and quality of the material model and thereby also the amount of calibration tests needed depend on the problem at hand. The basic unit cases used for calibration of the stress–strain relations are tension/compression tests at a constant strain rate and for a range of temperatures. The temperature is held constant during each test. The inclusion of phase change effects requires these tests to be repeated for the possible microstructure that can

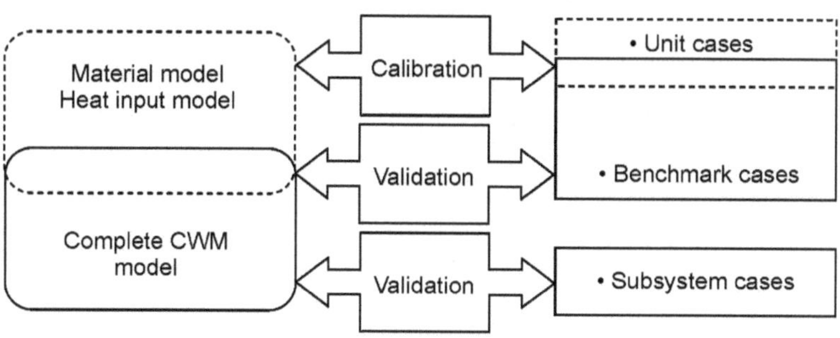

8.5 Proposed calibration and validation strategy.

Table 8.3 Material tests for calibrating strain–stress behaviour

Tests at a set of constant temperatures.
Varying strain rates for each of the tests at higher temperatures where viscous effects become significant.
Each of the tests performed at varying microstructures, typically different phase compositions. A given microstructure is created by subjecting the test piece to a temperature history before the mechanical testing.

occur. The tests may also be done over a varying range of strain rates if rate-dependent plasticity is to be accounted for. The general set of mechanical tests is listed in Table 8.3.

It is recommended to record time, elongation, diameter change and force as well as temperature during the tests. It is possible to obtain quite uniform conditions in the specimen so that accurate, true strain–stress data can be obtained from the measurements. Non-uniform conditions are unavoidable for large strains or large strain rates. Typically, neither is needed in CWM models. If the conditions are non-uniform, then a finite element model of the whole test piece is needed in the calibration process as in Berglund and Alberg [72] and Kajberg *et al.* [220]. More details about calibration are given in Section 8.5.3.

Free expansion (dilatometry) tests where length changes are measured during thermal cycling are used to calibrate the thermal expansion. These tests also give the information that will be needed if microstructure models are to be used [61, 221] in a TMM model of the process.

Inclusion of the transformation plasticity effect when martensite or bainite are formed is often based on the model proposed by Leblond (see Section 9.7.2). This model does not bring in any new material parameters. It is often used without separate calibration or validation. Otherwise a typical test is to subject a specimen during cooling for a given stress during the phase change [222–224] (see Fig. 8.6). The large difference between the two curves in Fig. 8.6 is due to the transformation plasticity. Otsuka *et al.* [225] propose the use of a four-point bending test to calibrate this part of the model.

Two sets of tests are proposed for validation of the material model. The primary one is the Satoh test [66] as it gives a thermomechanical loading resembling the material in the weld line. The test is described in Section 4.2. A complementary validation test is used if strain rate effects are accounted for in the material model.

A sequence of tests with varying strain rates, at different temperatures and for varying microstructures, were used to calibrated a rate-dependent model for a Greek Ascoloy material [216, 226]. Strain rate jump tests at varying temperatures and microstructures were used for validation of the rate dependency. One example is shown in Fig. 8.7. The Satoh test is useful for validation of the material model as it resembles the welding cycle (Fig. 8.8). The case had a

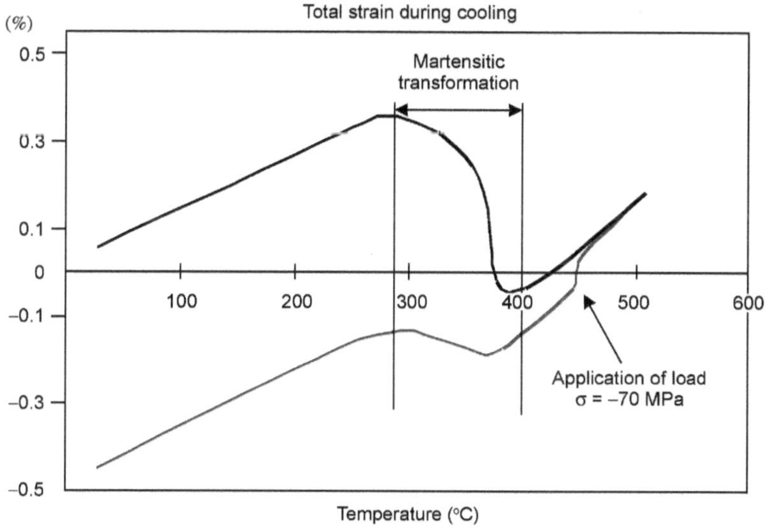

8.6 Transformation plasticity due to applied stress during martensite formation [222].

heating rate of 50 °C/s and a cooling rate of 10 °C/s. A finite element model for the whole Satoh test specimen was used [72] in the calibration procedure (see Fig. 8.9). The computed data shown in Fig. 8.8 is performed with one element using the CWM_Lab software.

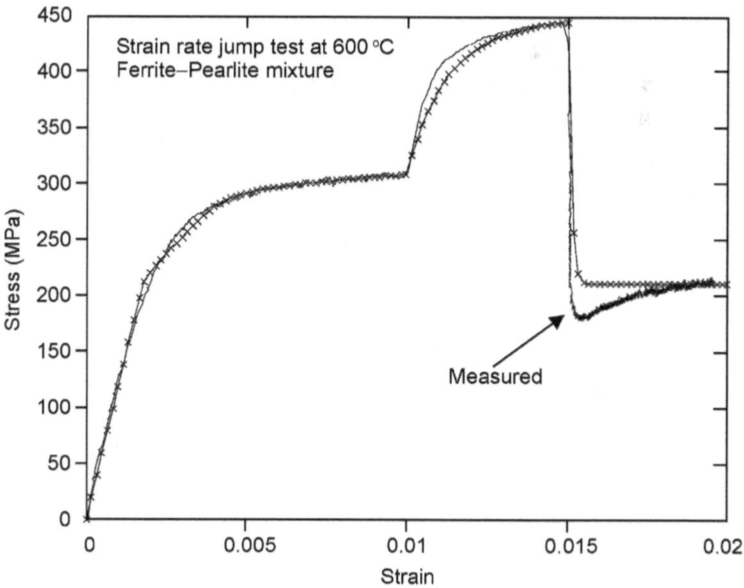

8.7 Strain rate jump test used for validation of rate dependency.

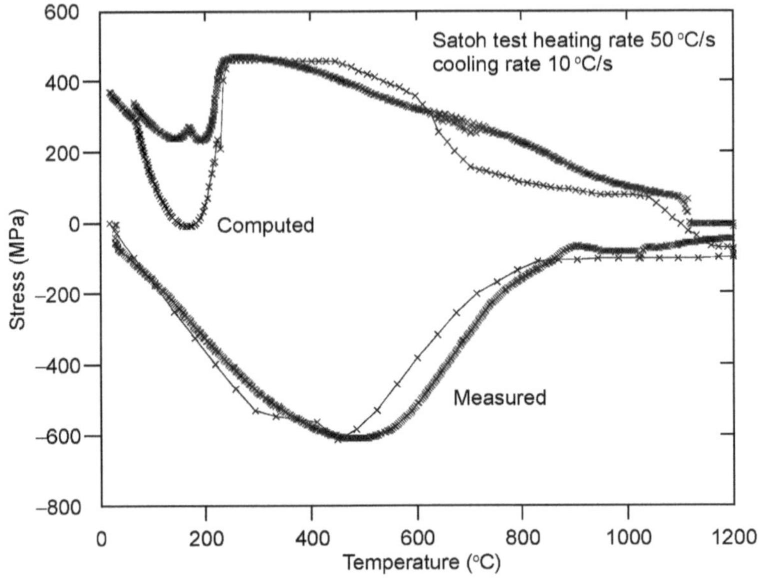

8.8 Satoh test used in validation of material model.

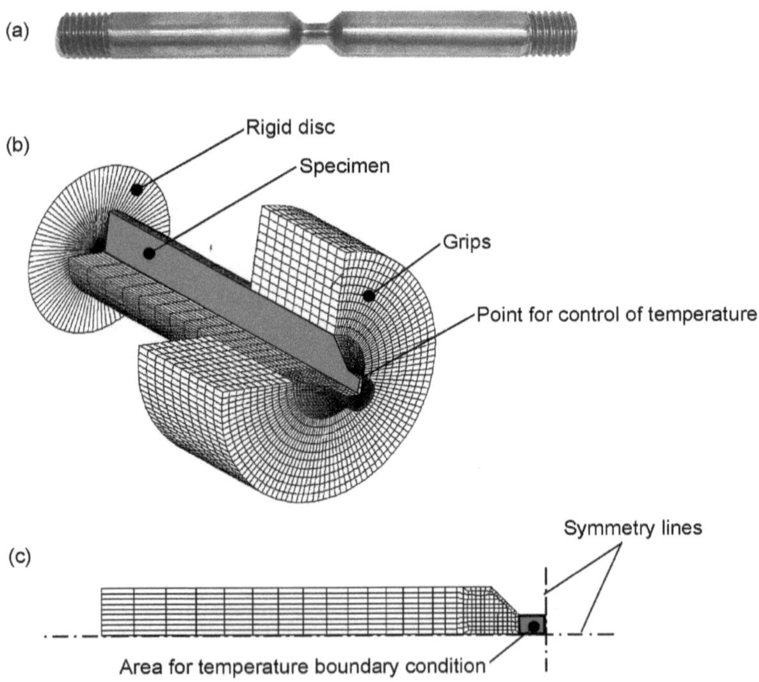

8.9 (a) Satoh test piece and (b), (c) axisymmetic finite element model [72].

8.5.2 Calibration and validation of heat input model

A lot of work has been done regarding the calibration of the heat input model as this is an important issue. However, it should be noted that in many cases it is sufficient to input the correct amount of heat in approximately the correct time and space. The details of the weld pool temperature distribution are less important and this zone is simplified considerably in classical CWM (see Section 9.12). Many published papers show a good agreement between computed and measured temperatures after the calibration process. The calibration of the heat input model can be done on the benchmark test. However, it can be performed on any geometry provided the welding procedure, joint preparation and the material correspond to the final application. Some benchmark cases are described in Section 8.5.4.

The measured cross-section of weld metal and/or HAZ can be used to calibrate the heat input model. The shape in the welding direction cannot be observed. However, an idea of the length of the weld pool and the welding speed is often sufficient. The cross-section dimensions are important as the heat flow in the transverse direction dominates in welding. The notion behind this approach is that if the weld pool boundary is correct, then the temperature field outside this region will also be correct. This was found to be the case in Lundbäck *et al.* [227]. It requires that all thermal properties are correct for all temperatures. Furthermore, the weld pool geometry is not only a question about its cross-section. Lundbäck and Runnemalm [17] calibrated the finite element by a weld process model by Sudnik *et al.* [228]. This enabled the comparison of the longitudinal shape of the weld pool also (Fig. 8.10).

Transient measured temperatures near the weld are often used to calibrate the heat source. However, the best approach would be to use the approach above for calibration and then use transient temperatures for validation of the thermal model. Very many of the papers with measurements listed in Lindgren [209] are concerned with the temperature field. Transient temperatures are commonly measured with thermocouples. However, infrared thermography can also be used, as by Ortega *et al.* [229].

A 28-pass weld was simulated by Lindgren *et al.* [3] with a simple approach to add heat input by prescribing the elements corresponding to the filler material to have the melting temperature T_{liquidus} during a time estimated from the length of the weld pool and weld speed. No further calibration was performed. An example of measured and computed temperatures is shown in Fig. 8.11.

If no separate measurements are performed to validate the thermal model, then the validation of the mechanical behaviour of the model, described in Section 8.5.4, can be considered as a validation of the thermal model also, as this is the load that drives the measured mechanical deformation.

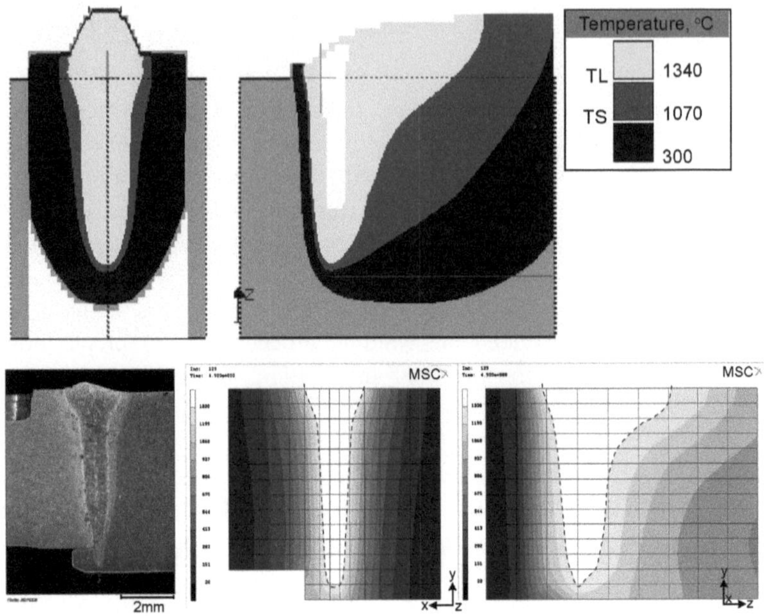

8.10 Weld pool calibrated using weld process model [17].

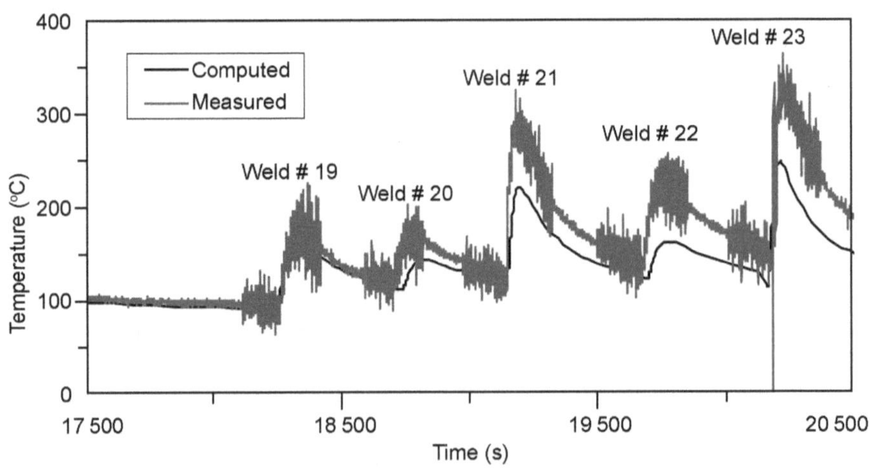

8.11 Measured and computed temperatures for multipass weld [3]. The scatter in measurements is due to disturbances from the electric arc.

8.5.3 Calibration procedure

Calibration of models is a general procedure. Many models for varying phenomena contain parameters that need be determined. Parameter identification can sometimes be straightforward, as individual parameters can be directly identified from tests. Typically the heat input model is in most papers determined by a manual trial and error procedure. However, this is not possible for models where parameters are not easy to identify directly by some features in the test results. Then a systematic and objective computer-based procedure for parameter identification is preferable. This is formulated as a maximisation of a likelihood function or a minimisation of an error [230, 231]. This is often preferable when calibrating the material model. An error measure is minimised with respect to the material parameters. The calibration of the material parameters, in the vector $\boldsymbol{p}$, is formulated as

$$\text{Minimise } e(\boldsymbol{p}) \qquad \boldsymbol{p} \in R^n \tag{8.1}$$

Subject to constraints

$$C_{\mathrm{i}}^{\mathrm{eq}}(\boldsymbol{p}) = 0 \qquad i = 1 \ldots n_{\mathrm{eq}}$$

$$c_{\mathrm{i}}^{\mathrm{in}}(\boldsymbol{p}) \leq 0 \qquad i = 1 \ldots n_{\mathrm{in}}$$

$$\boldsymbol{L} \leq \boldsymbol{p} \leq \boldsymbol{U}$$

where n is the number of unknown parameters and e is an error measure. The constraints can be nonlinear equations, n_{eq} is the number of equality and n_{in} the number of inequality constraints. $\boldsymbol{L}$ and $\boldsymbol{U}$ are lower and upper limits for the parameters.

Bruhns and Anding [231] use the error measure

$$e(\boldsymbol{p}) = \frac{1}{2}\boldsymbol{r}^{\mathrm{T}}\boldsymbol{G}\boldsymbol{r} \tag{8.2}$$

where $\boldsymbol{G}$ is a diagonal matrix with individual standard deviation errors for the measured data. The measured data are split into independent, driving, variables and dependent variables. The first are fed into the model and the latter are the output from the model. The vector $\boldsymbol{r}$ holds the square of the difference between computed and experimentally obtained values. The length of this vector is often lower than the number of sampling points in the measurements as these can sometimes be numerous. Sometimes a combination of smoothing and data reduction procedures are applied to reduce scatter and the amount of data in the measurements [232].

Mahnken and Stein [233] discuss the possibility of adding a regularisation term to Eq. [8.1] in order to overcome the problem of numerical instabilities due to non-uniqueness. They also discuss the use of the eigenvalues of the Hessian matrix in order to evaluate the robustness of the solution. The matrix $\boldsymbol{G}$ can be used to make the error nondimensional [234] and include user-defined weight functions, $\boldsymbol{w}$, to include weighting of the data for different purposes:

$$G_{ii} = w_i \Delta_i \qquad \text{(no sum over } i\text{)} \tag{8.3}$$

where w_i is a weight factor evaluated at sampling point number i associated with an interval Δ_i. This can be used to put equal weight on all sampling points by $w_i = 1$ in the case where the length of the sampling interval, Δ_i, varies during the test.

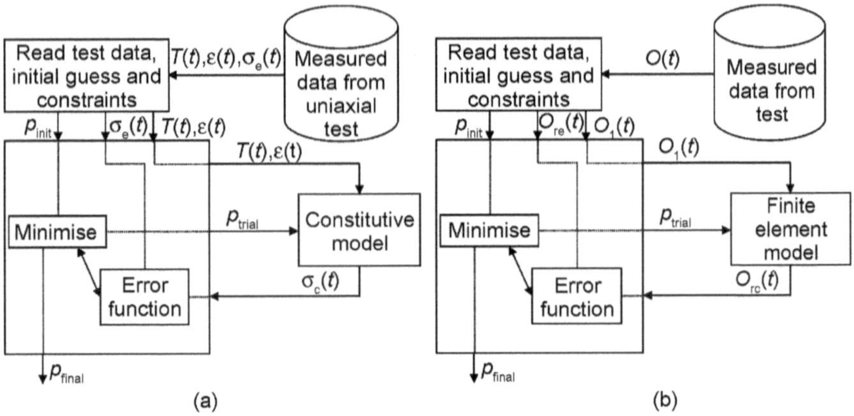

8.12 Parameter fitting with homogeneous test giving stress and strain data (a) and with tests where inverse modelling is needed (b).

A typical approach in constitutive models is shown in Fig. 8.12 where the stress is the dependent variable, giving

$$r_i = (\sigma_c - \sigma_e)_i^2 \qquad [8.4]$$

where subscript i denotes the sampling at a given time or strain. A direct parameter fitting can be applied for tests where it is possible to measure over a homogeneous deformed volume in order to obtain strain and stress (Fig. 8.12a). p_{final} is the material parameters that minimise the difference between computed σ_c and measured stress σ_e. A more general approach is to use an inverse modelling technique [234–239] where a finite element model of the test is used (Fig. 8.12b). O are measured quantities and they are separated into loading (independent quantities), O_l, and experimental results (dependent quantities), O_{re}. The independent values are prescribed to the finite element model and the computed results O_{rc} are compared with the measured in order to find the best material parameters.

The optimisation procedure applied to Eq. [8.1] can be either a deterministic or stochastic method [230]. Typically some kind of gradient method is used in the first approach. These methods can be quite efficient but find only local minima [240, 241]. Thus they are very dependent on the starting value. Genetic algorithms can be considered as stochastic algorithms [242, 243]. They have an underlying logic of survival of the fittest but also use random generation when creating new candidates for the parameters needed. They are quite independent on the starting values but require more iterations. Thus they can be combined with gradient methods to obtain good starting values [231]. The system developed by Furukawa *et al.* [244] combines these methods with gradient-incorporated continuous evolutionary algorithms (GICEA). Bruhns and Anding [231] also used parallel computing to speed up the procedure. Huber and Tsakmakis [245] used a neural network for parameter fitting.

8.5.4 Validation of model using benchmark cases

The benchmark is a special fabricated case. The boundary conditions should be as well defined as possible and the measurements carefully done. Several experiments on nominally the same benchmark will give information about scatter in measurements as well as the robustness of the case. Unstable or not robust cases are sensitive to small changes in initial data or boundary conditions and therefore not suitable as benchmark tests. These issues are discussed in Chapter 11. The thermal model is usually only validated in this test as discussed in Section 8.5.2. The proposed calibration and validation strategy is complete with the benchmark cases. The exclusion of a subsystem case from the strategy does not mean that it would not be useful: however, the idea is to have benchmark cases that capture the main characteristic of the final application, so that the validation based on them is sufficient to give confidence in the developed model for the intended application.

Different variants of welded plates or circular discs have been common benchmark cases in the scientific literature. One example of the geometry for a benchmark case is shown in Fig. 8.13. It resembles the early butt-welded cases used by Andersson [88] and Jonsson *et al.* [71] where extensive measurements were done. However, their dimensions and the heat input were quite different and therefore the out-of-plane displacements were relatively smaller in their cases. The geometry in Fig. 8.13 has been used in several investigations at Luleå University of Technology [216, 246] supervised by Lindgren. It was been found [72] that the use of a flat plate was not a good option: the behaviour is not robust

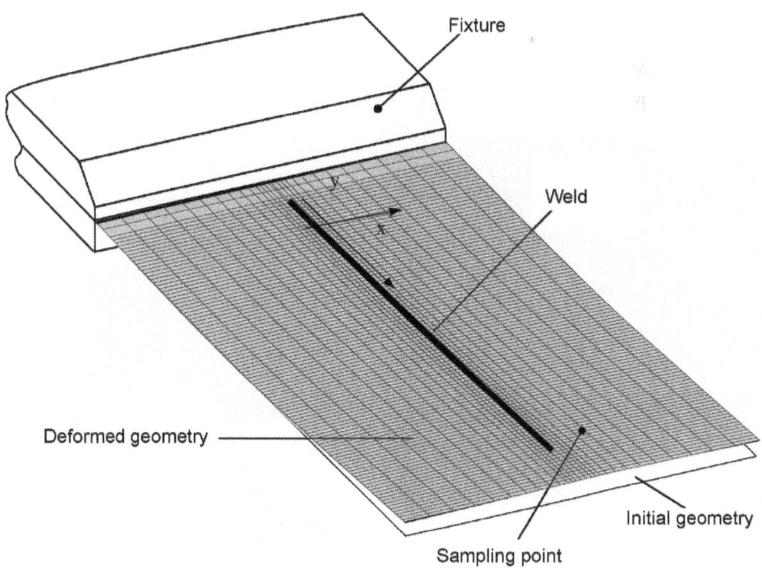

8.13 Benchmark case for validation.

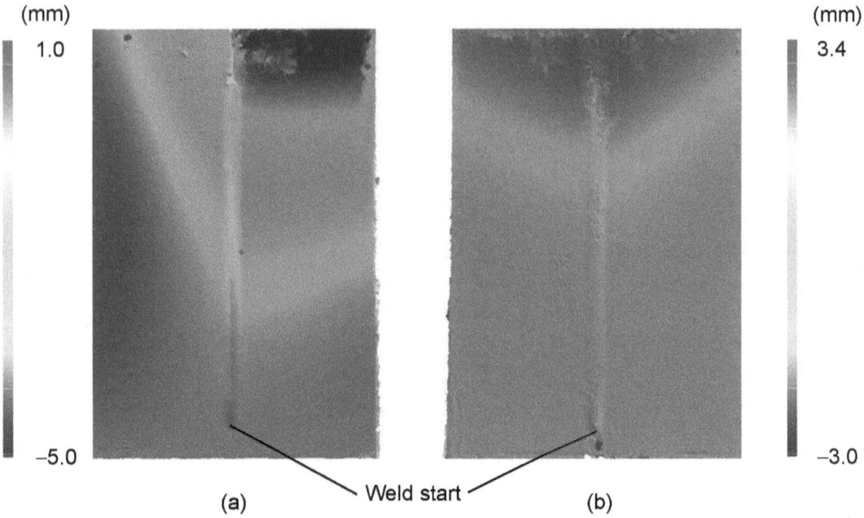

8.14 Measured field of deflection of two nominally identical subsystems with 'flat' plates [72].

since buckling may occur and whether the plate bends upwards or downwards depends on details on the initial geometry of the nominally 'flat' plate. Two examples of measurements on nominally the same plate are shown in Fig. 8.14. Including the exact initial geometry into the finite element model gave a good agreement with measurements (Fig. 8.15). Introducing an angle between the plates (Fig. 8.16) makes the deformation behaviour more robust. The table in the figure shows the final deflection of the end of the plate versus the angle between

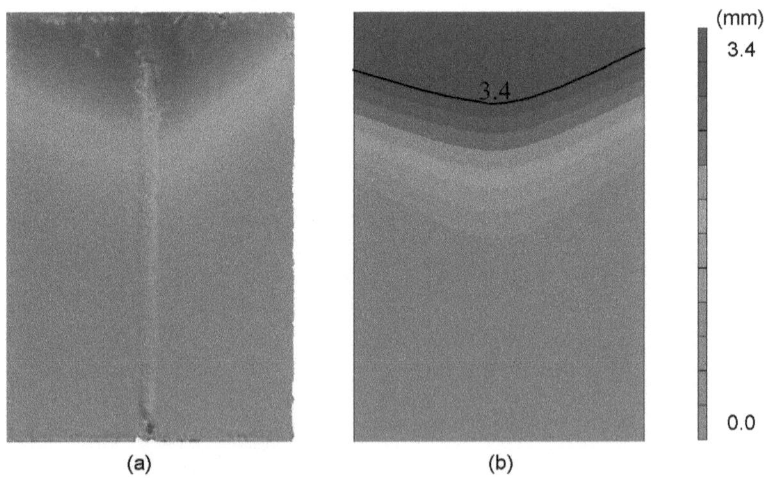

8.15 Out of plane deformation after welding. (a) Measured out of plane deformation. (b) Simulated out of plane deformation [72].

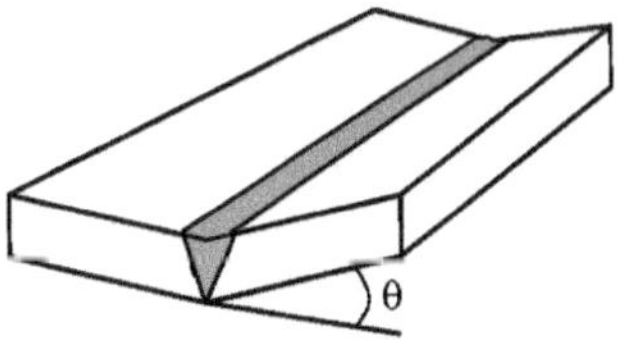

Model	θ(°)	Displacement (mm)
WM 1	–	4.3
WM 2	1	5.7
WM 3	−1	5.8

8.16 Use of initial butterfly angle to create a robust deformation behaviour.

the two halves of the plate. Therefore, an initial angle of 1° has been used in the subsystem case and the clamping was manufactured in order to be consistent with this angle.

Another problem is to ascertain whether the clamping can be modelled as perfectly rigid, especially if the plate is thick. It is believed in the current case that the clamping is rigid as the plate is only 1.7 mm thick and therefore it has not been included in the model but replaced by fixed displacements in the model.

The purpose of this configuration of the benchmark is to have a case that is not too rigidly fixed. This is because the CWM models used at Volvo Aero Corporation (see Chapter 13) are primarily used to minimise the welding deformations. The residual deformations are more difficult to predict than residual stresses for these cases. Residual stresses are more robust in the sense that errors in deformations are less visible on the largest residual stresses as these are usual at the yield limit of the material. Then the effect of the error in the displacement field on the strains and, from these, on these residual stresses is small as the hardening modulus is low compared with the modulus of elasticity. Therefore, a relatively flexible case is used as this will reveal deficiencies in the model better than a rigid case. Several publications have revealed problems when attempting to predict deformations of similar cases. They require more accurate models. This is elaborated further in the discussion about modelling issues in Chapter 9.

Transient temperatures and displacements at several locations as well as residual stresses and deformations field have been measured [216, 246]. The overall agreement is good for most variants of the models used. The focus is on the vertical deflection of the end of the plate as this is the most difficult part of the validation data. The current state of this work is still in progress. A number of models with varying degrees of complexity and also with shell or solid elements have been used in simulations of this problem. Residual stresses have been measured and compare well with computed values but the deflection of the end of the plate is still not sufficiently accurate (Fig. 8.17).

8.6 Validation using subsystems and complete systems

A subsystem is a case that can be identified geometrically as a part of the complete system. There are no publications known to the author where both

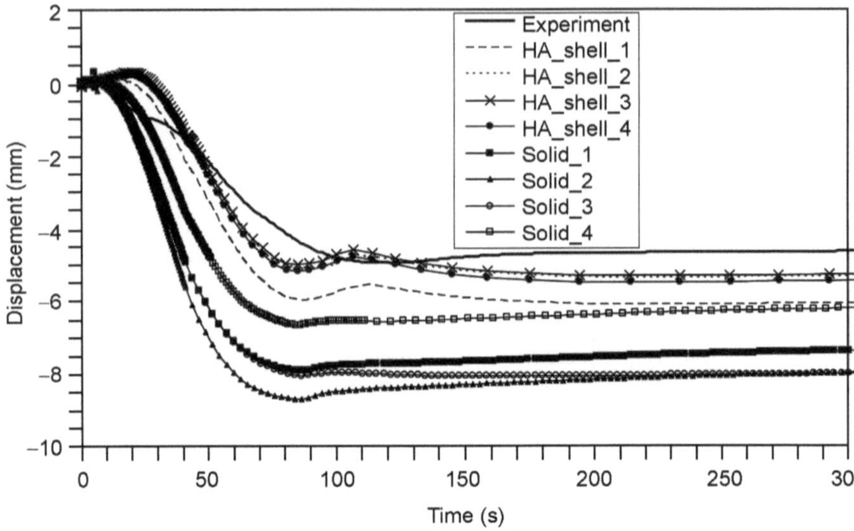

8.17 Measured and computed deflection of plate at a point at the end of the plate and near the weld.

subsystems and complete systems have been used in the validation. Typical papers validate their models on a subsystem, such as a plate representing a part of a boiler vessel, or they validate on the welded component itself, such as a welded pipe or plate.

9

Modelling options in computational welding mechanics (CWM)

Several options when modelling welding, within the context of classical CWM models, are described below. The most important modelling considerations are concerned with material and heat input models, and choice of type of geometry together with the discretisation of the problem. The heat input model is specific for CWM models whereas the others are common for all types of applications. The models are described together with references to research papers where they have been applied. The chapter is a background to the next chapter where a modelling strategy is proposed.

9.1 A note about computability in CWM

It should be noted that some welding configurations might be less robust and even prone to instability. Then the computability [217] is low and the guidelines in the following may be inadequate for setting up accurate models. Stone *et al.* [73], Lundbäck and Runnemalm [17] and Pilipenko [70] had problems with agreement between measured and computed deformations of thin plates. Some of these cases might have been due to buckling tendencies. However, from an engineering point of view, it is not important for the prediction of the behaviour in these cases to be quantitatively correct. Then it is sufficient to be able to discover this phenomenon, as it is necessary to redesign the welding process in order to make it robust. Chapter 11 discusses the use of simulations to evaluate the robustness and stability of a welding process.

9.2 The importance of material modelling

The material modelling process involves several issues:

- Determine all relevant phenomena, i.e. the scope of the material model.
- Perform calibration tests.
- Choose material model.
- Determine model parameters, discussed in Section 8.5.1.
- Validate material model, discussed in Section 8.5.1.

Usually there is a need to iterate through the process above with complementary tests to determine all the parameters in the model uniquely or to change the model so it captures all the relevant phenomena discovered during the process. It should be noted that there is a large wealth of information about models for different materials under varying conditions available in research papers and books (e.g. [62, 120, 247–254]).

The material modelling, together with the uncertain net heat input, is one of the major problems in simulation of welding. McDill *et al.* [165] investigated the relative importance of thermal and mechanical properties of stainless and carbon steels in welding simulations. Two bars of dimensions 20 in. × 2 in. × 0.5 in. ($500 \times 50 \times 12\,\text{mm}^3$) were 'welded' along the free edge. No fixture was used. One bar was made of a stainless steel and the other of a carbon steel. The final radius of curvature obtained was compared with computed values where the material properties were taken as those for the carbon steel (MS) or the stainless steel (SS). The results are shown in Table 9.1. The somewhat unexpected conclusion was that thermal properties play a more important role than mechanical properties in explaining the different behaviours of these steels. This is because thermal dilatation is the driving force in the deformation and the bar was free to bend. Thermal dilatation is determined by the temperature field and is therefore strongly influenced by the thermal properties. The thermal analysis is in general straightforward compared with the mechanical analysis. It entails few numerical problems, with the exception of the large latent heat during the solid–liquid transition, and it is easier to obtain thermal properties than mechanical properties of a solid. Therefore, the smaller space devoted to thermal properties in the following should not be taken as an indication that they are less important.

A large range of material and surface properties may be needed in CWM models. A short overview is given in Table 9.2, which also shows what material properties are discussed in the following. The high-temperature mechanical behaviour is modelled in an approximate way due to several factors: experimental data are scarce, material that is too soft causes numerical problems [170, 255] and any approximations introduced do not greatly influence the final

Table 9.1 Curvature for combinations of thermal and mechanical properties [165]

Test	Thermal properties	Mechanical properties	Radius of curvature [m^{-1}]
A	MS	MS	41.6
B	SS	SS	11.9
C	MS	SS	29.6
D	SS	MS	16.3
Experiment	MS	MS	23.3
Experiment	SS	SS	8.1

Table 9.2 Different material and surface properties

Bulk properties	Surface properties
Density, Section 9.4	Boundary properties, Section 9.11 convection radiation
Thermal properties, Section 9.5 heat conduction heat capacity latent heats	
Mechanical properties Elasticity, Section 9.6 Plasticity, Section 9.7	Interface properties, not discussed contact resistance friction heat due to friction
Thermomechanical properties, Section 9.8 thermal expansion heat due to plastic dissipation	
Microstructure evolution, Section 9.9	

residual stresses. Many analyses use a *cut-off temperature* above which no changes in the mechanical material properties are accounted for. It serves as an upper limit of the temperature in the mechanical analysis. The meaning of using a cut-off temperature may vary as some studies only apply this cut-off to some properties. Ueda and Yamakawa [160, 161] did not heat the material higher than 600 °C and Fujita *et al.* [163] limited the maximum temperature to 500 °C. Ueda *et al.* [256] assumed that the material did not have any stiffness above 700 °C. They called this the mechanical rigidity recovery temperature above which Young's modulus was set to zero. It is therefore likely that this corresponds to the cut-off temperature. Hepworth [255] used 800 °C as a cut-off temperature above which he did not include any thermal dilatation. Free and Porter Goff [257] completely ignored the temperature dependency of Young's modulus, neglected hardening and used 900 °C as a cut-off temperature. Furthermore, they only followed the cooling phase of the compiled temperature envelope. Tekriwal and Mazumder [258] varied the cut-off temperature from 600 °C up to the melting temperature. The residual transverse stress was overestimated by 2–15% when the cut-off temperature was lowered.

9.3 Effect of temperature and microstructure

The complete thermomechanical history of a material will influence its material properties. However, this can be approximated to a dependency only on the current temperature and deformation for many materials. The effect of phase changes on thermal and mechanical properties may be summarised as in Table

Table 9.3 Effects of phase changes

Memory of structure, Section 9.3	Volume changes, Section 9.3
Transformation plasticity, Section 9.7.2	Latent heats, Section 9.5

9.3. The table does not include the fundamental effect that current properties depend on the current microstructure. It lists the direct effects due to the *changes* in phase composition. The indirect effect is that the changed microstructure changes the properties.

The simplest and most common approach is to ignore the microstructure change and assume that the material properties depend only on temperature. The effect of phase changes may be ignored for austenitic steels [191, 192, 259–261], copper [136] and Inconel [125, 179, 262, 263]. But phase changes have also been ignored in the case of ferritic steels [3, 127, 160, 172, 180, 264, 265]. Then, for example, the thermal dilatation is the same during heating and cooling as in Fig. 9.1.

A need to account for the phase transformations for ferritic steels became apparent. The effect of the phase transformations on the thermal dilatation was included first by Ueda and Yamakawa [161] and later by Ueda *et al.* [266, 267]. Andersson [88] also accounted for this effect. The approach used by Andersson (and also in [71, 75, 171, 268]) is based on given property–temperature curves. Different curves are chosen in the analysis depending on some characteristics of the temperature history at the considered point in the model. Usually these characteristics are the peak temperature and the cooling rate between 800 and 500 °C. They are the primary parameters that determine the obtained microstructure of steels. The cooling rate is approximately the same in the whole HAZ. Therefore different property–temperature curves are chosen during the cooling phase depending on the peak temperatures, as shown in Fig. 9.2 in the case of thermal dilatation. A single, common curve is used during the heating

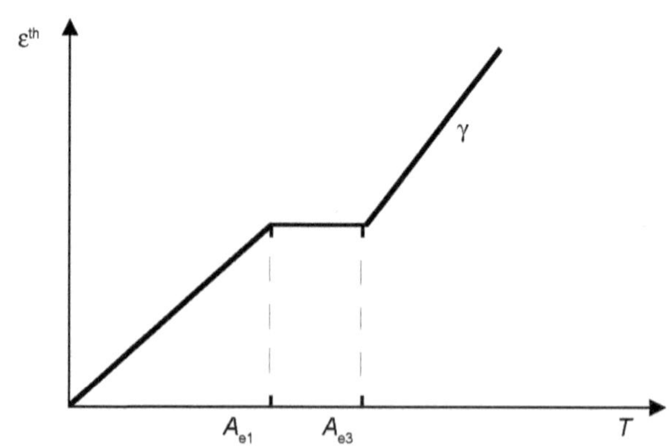

9.1 Temperature-dependent thermal dilatation ignoring phase changes.

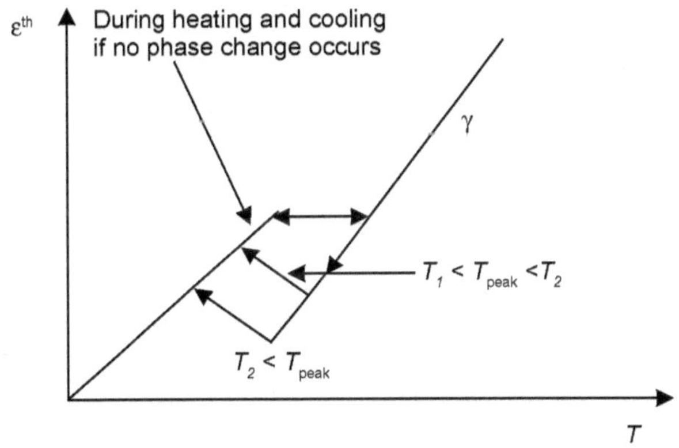

9.2 Temperature-dependent thermal dilatation accounting for phase changes in a simplified manner where data during cooling depend on peak temperature.

phase. Different curves are chosen during cooling depending on the value of the peak temperature, T_{peak}. It is also possible to interpolate between these curves with respect to the peak temperature instead of just switching between them.

The most flexible way to include the effect of the temperature history is to compute the evolution of the microstructure in the material. Each phase is assigned temperature-dependent properties, and simple mixture rules are used to obtain the macroscopic material properties. This coupling of thermal, metallurgical and mechanical (TMM) models has been used [37, 65, 174, 269–281]. An example of a mixture rule is shown in Fig. 9.3 where the thermal expansion coefficient is given one constant value for austenite and another value for the other phases. The difference in specific volume between the austenite and its

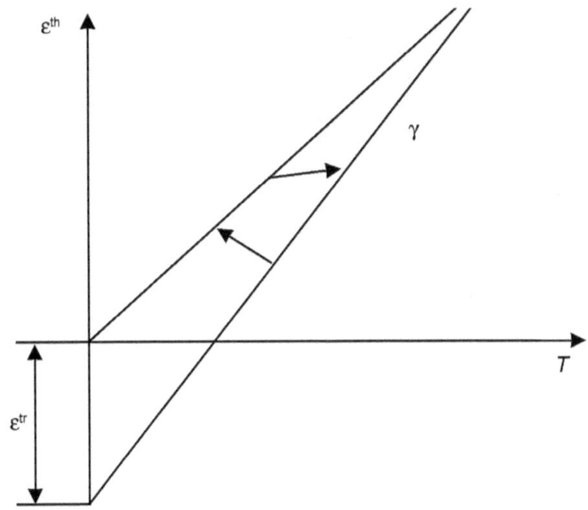

9.3 Temperature-dependent thermal dilatation in a thermo-metallurgical-mechanical model.

decomposition products is denoted by ϵ^{tr} which is this difference extrapolated to 0 °C in this case. This kind of approach uses models for the microstructure evolution as discussed in Section 9.9.

The TMM approach where a model for the microstructure evolution is combined with constitutive models for each phase requires some mixture rules. It is also necessary to have a rule for how the structure of the disappearing phase affects the new phase. Most papers have used linear mixture rules for all material parameters.

$$p = \sum X_i p_i \qquad [9.1]$$

where X_i is the volume fraction of phase i and p_i is the property of the phase at the current temperature. p is the macroscopic, average, property for the material. This mixture rule can be applied to the thermal dilatation for the two-phase case illustrated in Fig. 9.3. The difference between the slopes of the curves gives the difference in thermal expansion coefficient for each phase. The shift of the austenite curve downwards is due to the smaller specific volume of austenite. Thus the linear mixture rule includes this effect. Sjöström used a linear mixture rule for simulating hardening of steels [282].

The problem of determining an appropriate mixture rule for plastic properties is much less straightforward than for thermal dilatation, where it works well. The macroscopic flow stress will depend not only on the fraction of the phases but also on their distribution. However, the density type of microstructure models used in CWM do not contain this information. The common choice is to use a linear mixture rule for the flow stress. It is written as

$$\sigma^y = X_1\sigma_1^y + X_2\sigma_2^y = (1 - X_2)\sigma_1^y + X_2\sigma_2^y \qquad [9.2]$$

for a two-phase mixture.

Furthermore, a relation between the macroscopic hardening parameter, usually denoted by the effective plastic strain, and its corresponding quantities in the phases is needed. This parameter represents the internal dislocation structure created during the deformation that causes the hardening. It can be assumed that the effective plastic strain develops equally in all phases, the iso-strain assumption. However, this is less likely when the phases have large differences in their yield stress. Another approach is to take into account that less deformation takes place in the harder phase by the use of the iso-work principle [283]. This can be written as

$$\sigma_1^y(\bar{\epsilon}_1^p)\dot{\bar{\epsilon}}_1^p = \sigma_2^y(\bar{\epsilon}_2^p)\dot{\bar{\epsilon}}_2^p \qquad [9.3]$$

where $\sigma_1^y(\bar{\epsilon}_1^p)$ denotes that the flow stress of phase 1, σ_1^y, depends on the current value of the plastic strain, $\bar{\epsilon}_1^p$, in this phase. This is used together with the mixture rule giving the macroscopic plastic strain

$$\dot{\bar{\epsilon}}^p = (1 - X_M)\dot{\bar{\epsilon}}_\gamma + X_M\dot{\bar{\epsilon}}_M \qquad [9.4]$$

Relations [9.1]–[9.4] lead to additional considerations when computing the macroscopic hardening modulus of a material, which is defined as

$$H' = \frac{d\sigma^{y}}{d\bar{\epsilon}^{p}} \quad [9.5]$$

Other mixture rules have been used for the flow stress. A power law mixture rule was proposed by Gladman *et al.* [284]:

$$\sigma^{y} = \left(1 - X_2^{N}\right)\sigma_1^{y} + X_2^{N}\sigma_2^{y} \quad [9.6]$$

where N is a calibration parameter. Bouquerel *et al.* [285] found the value $N = 2$ fit their data.

Leblond *et al.* [286] propose a non-linear mixture rule of the type

$$\sigma^{y} = [1 - f(X_2)]\sigma_1^{y} + f(X_2)\sigma_2^{y} \quad [9.7]$$

which is obtained from numerical experiments for an austenite/martensite mixture. Bouaziz and Buessler [287] compared some mixture rules.

A rule of inheritance, or memory, of previous hardening is also needed during a phase change. How much of the dislocation density will be remembered after a phase transformation? There has been a general agreement to remove all accumulated hardening when a material melts [136, 288]. Mahin *et al.* [289] included the removal of plastic strains during melting as they had found that this was a source of discrepancy between simulations and measurements in their earlier work [290]. Ortega *et al.* [229] initialised all internal state variables to zero when the material melted. They also removed the deviatoric stresses creating a hydrostatic stress state in the weld pool, as did Dike *et al.* [186]. Friedman [126] and Papazoglou and Masubuchi [169] removed the accumulated plastic strains when the material melted and relieved the accumulated strains by multiplying the previously accumulated plastic strains with a factor due to solid state transformations. Devaux *et al.* [273] argued that it is reasonable that the memory of previous plastic deformation disappears for all solid state phase transformations in ferritic steels with perhaps the exception for the martensite formation. The latter involves very small displacements of the atoms during the transformations which they assumed did not affect the dislocations. However, the findings by Vincent *et al.* [69] do not support the assumption that the martensite inherits the hardening from the austenite phase. It may be that the inherited structure is completely erased during the martensite formation itself as the martensite has an initial high dislocation density. Brust *et al.* [291] and Dong *et al.* [190] introduced rate equations applied between an anneal temperature and the melting temperature for anneal strain. The introduction of these strains corresponded to the removal of plastic strains as they reduced the hardening. They also applied this to the elastic strains and thereby reduced the stress also.

9.4 Density

Simulation of welding can usually be performed as a quasi-static analysis (Section 3.3). The density is needed even for a quasi-static analysis as it is multiplied with the heat capacity in the thermal analysis. Handbooks may give the density as a function of temperature. This accounts for the volume change due to the thermal dilatation. However, one must know how the density is handled by the used finite element code. A constant density may be sufficient as input if the code itself computes the change in density when deformations are computed simultaneously with the temperatures, e.g. in a staggered approach. A temperature-dependent density may be used for a pure thermal analysis where no deformations are included.

9.5 Thermal properties

Fourier's law for heat conduction is the only model used in CWM. Then the solution of the heat conduction equation requires heat conductivity, λ, heat capacity, c, and the density. Density is discussed above. These thermal properties are temperature dependent, and may also depend on the temperature history as different phases may have different thermal properties. Furthermore, the latent heats due to phase changes are needed. Some of the earlier work, discussed in Lindgren [292], ignored the temperature variation of thermal properties. One example is Ueda and Yamakawa who used constant thermal properties in their early analyses [160, 161] but included the temperature dependency in later papers, e.g. [266, 267]. Goldak *et al.* [166] and Moore *et al.* [293] discuss and show the effect of using constant or varying thermal properties.

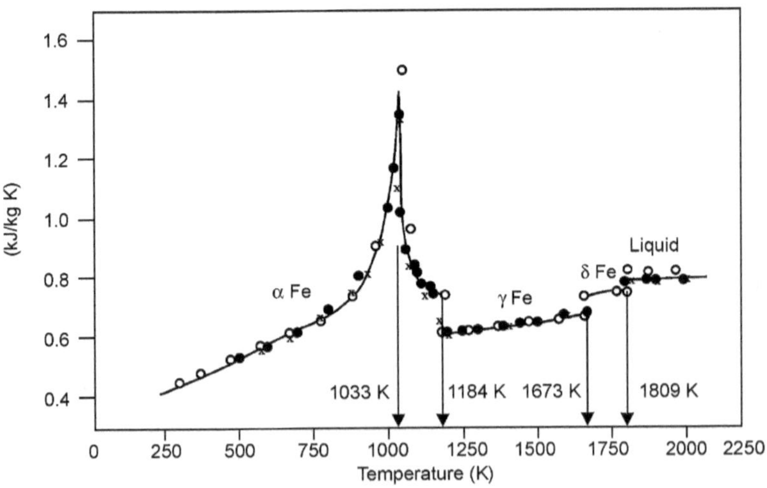

9.4 Heat capacity for pure iron, from Pehlke *et al.* [294].

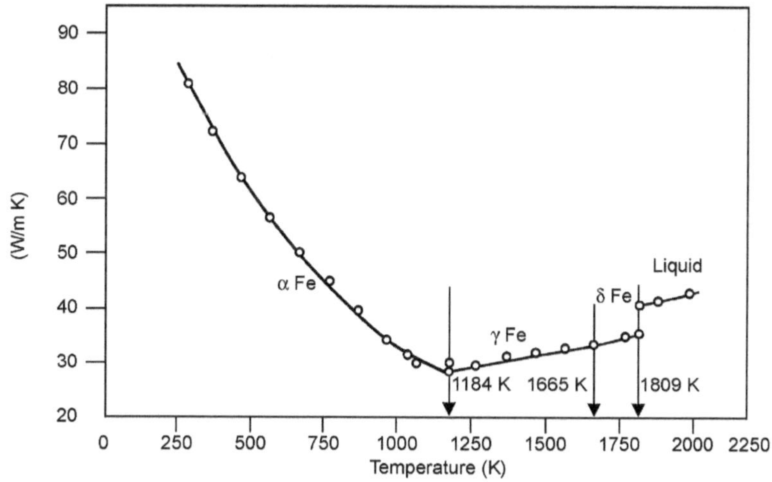

9.5 Heat conductivity for pure iron, from Pehlke *et al.* [294].

Thermal properties are more readily available or obtained than mechanical properties. This is not discussed here as many references with thermal properties can be found in Lindgren *et al.* [1]. The properties for pure iron are given in Figs 9.4 and 9.5 and for stainless steels in Fig. 9.6 and 9.7. An overview of these properties for different alloy steels is shown in Figs 9.8 and 9.9.

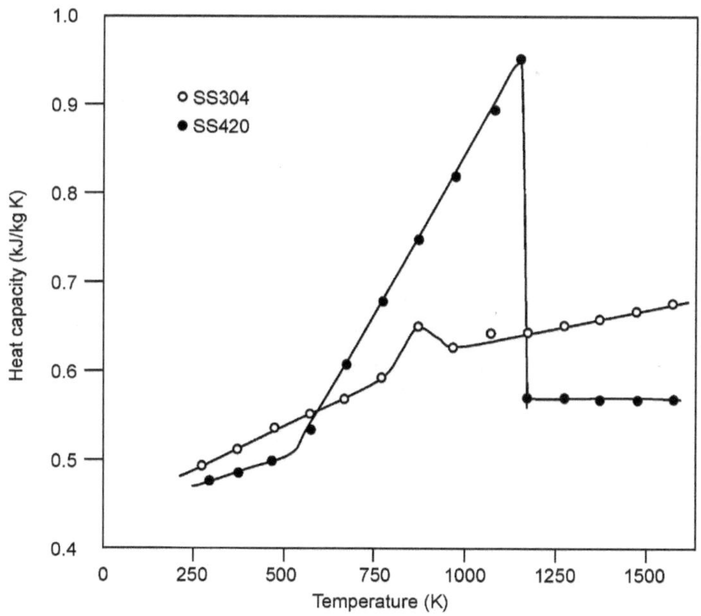

9.6 Heat capacity for two stainless steels, from Pehlke *et al.* [294].

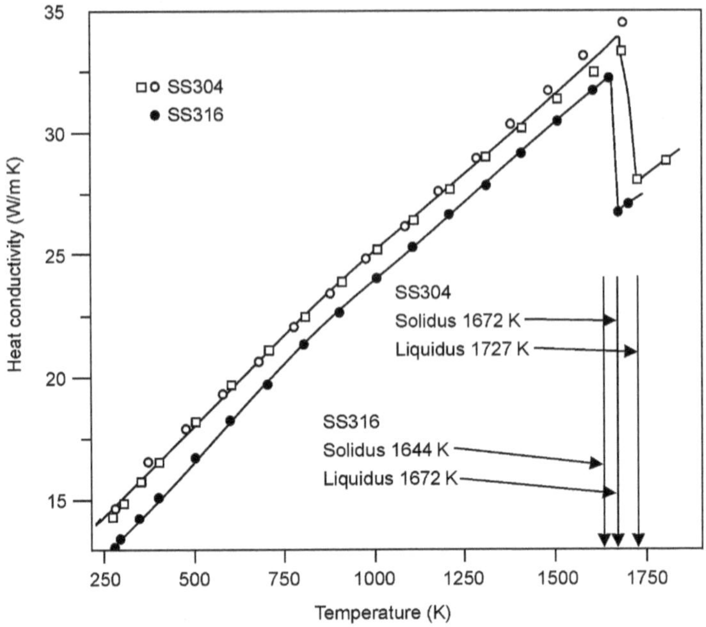

9.7 Heat conductivity for two stainless steels, from Pehlke *et al.* [294].

Latent heats due to the solid–solid phase changes are often ignored in many models and only the heat of fusion is included in the models. The effect of latent heats due to the solid phase transformation can be seen in, for example, Dubois *et al.* [295]. The magnitude of these latent heats can be estimated from diagrams in Pehlke *et al.* [294]. They used a latent heat of about 75 kJ/kg for the ferrite to austenite phase change in steels. Murthy *et al.* [77] used 92 kJ/kg for the austenite to pearlite phase change and 83 kJ/kg for the austenite to martensite

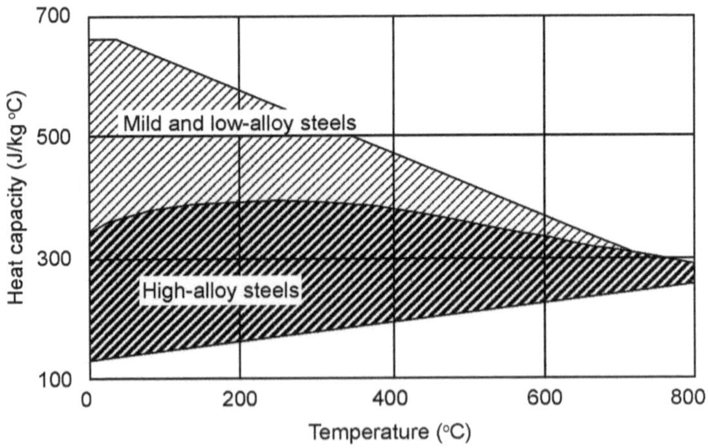

9.8 Heat capacity for different steels, from Richter [254].

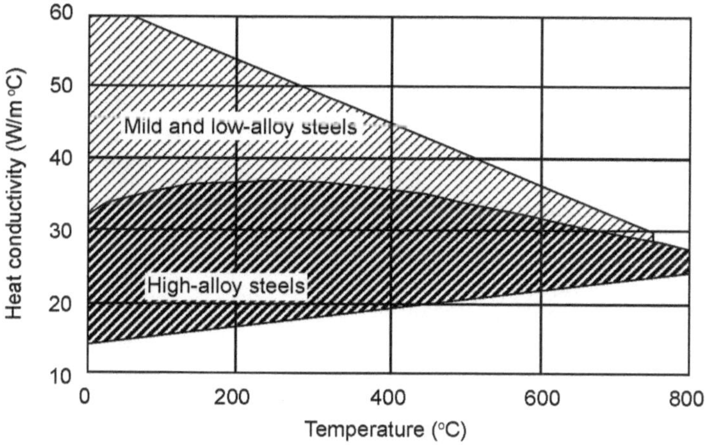

9.9 Heat conductivity for different steels, from Richter [254].

phase change. Latent heat due to melting/solidification is discussed in Section 9.10.

9.6 Elastic properties

All the models compute stress from elastic strains, although this may be done in different ways (see Sections 6.3 and 6.4). The properties needed for an isotropic material are the modulus of elasticity and Poisson's ratio. The latter has a smaller influence [258] on the residual stresses and deformations. An example of the modulus of elasticity is shown in Fig. 9.10 and Poisson's ratio is shown in Fig. 9.11. Several papers have assumed that Poisson's ratio goes towards 0.5 at the melting temperature as the molten metal may become an incompressible

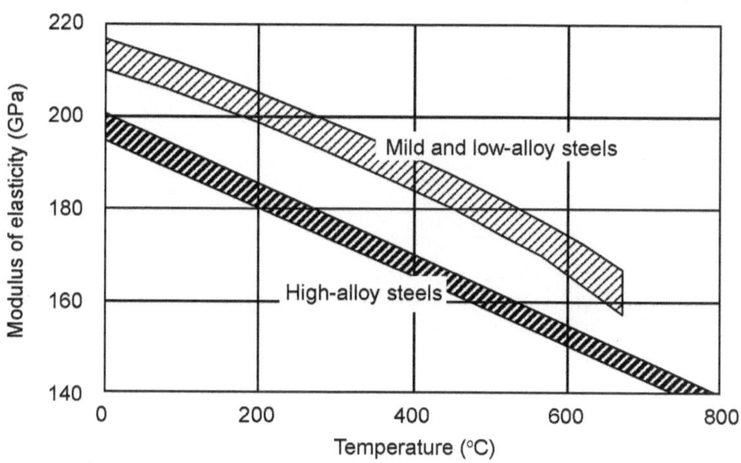

9.10 Modulus of elasticity for different steels, from Richter [254].

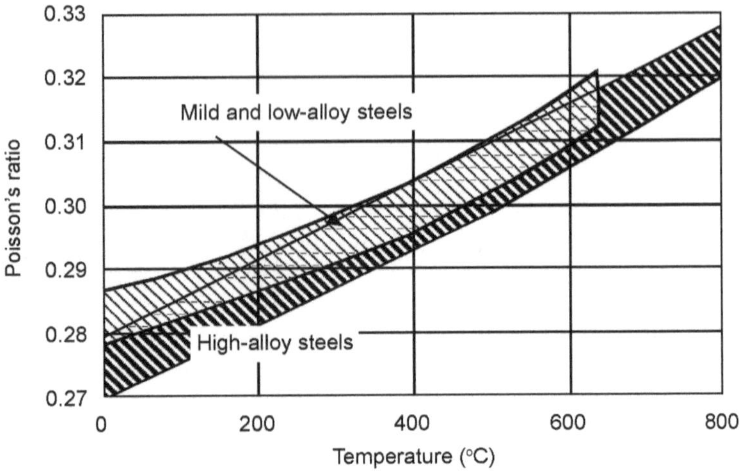

9.11 Poisson's ratio for different steels, from Richter [254].

fluid. However, there is no reason to assume that the solid phase just below the melting point has a Poisson's ratio near 0.5.

9.7 Plastic properties and models

The wide range of physical processes leading to inelastic deformations that can be active for varying stresses and temperatures makes it necessary to combine several models. The deformation mechanism map [118, 119, 296] is one way of illustrating this (see Fig. 9.12). This map is based on constant material structure

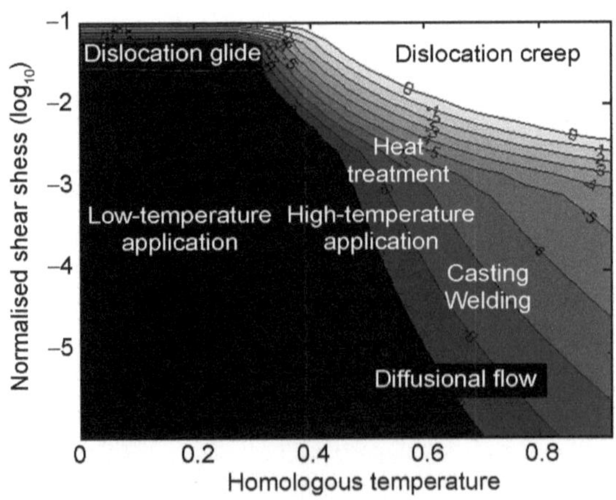

9.12 Deformation map for SS 316L stainless steel with grain size of 50 μm. Computed model based on Frost and Ashby [119]. Shear stress is normalised with the shear modulus at the corresponding temperature. The $\log_{10}$ for the steady state creep strain rates are given on the iso-curves.

and the strain rates shown by the iso-curves are those corresponding to the steady state creep rate. The discussion of models for the mechanical behaviour is structured by starting with temperature-dependent models where no phase changes occur or where they can be ignored. Thereafter different effects of phase changes on the mechanical material behaviour are discussed.

9.7.1 Plasticity and viscoplasticity models

The complex material behaviour in different temperature and strain rates is usually modelled by different models in different regions as shown Fig. 9.13. Rate-dependent plasticity or viscoplasticity accounts for rate effects and is more important at higher temperatures. However, the material experiences a high temperature during a relatively short time of the weld thermal cycle and therefore the rate dependency is often neglected. This was stated clearly by Hibbitt and Marcal [127] and others [266, 268] who included creep only when considering stress relief. Bru *et al.* [297], who studied the same problem as Roelens [174, 277], used tensile data for the strain rate of 0.1 s^{-1} and the rate-independent plasticity model. Sekhar *et al.* [298] showed the effect of using yield stress with temperature for two different strain rates. The difference is small for that particular case, except around 700–900 °C. Most studies in simulation of welding approximate the yield limit at higher temperatures and try not to make any elaborate adjustment for expected dominant strain rate, as the available data are scarce.

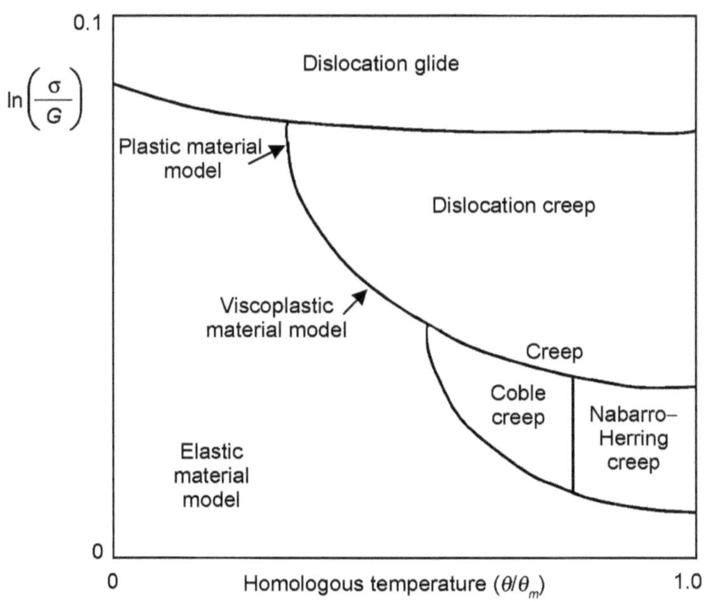

9.13 Deformation mechanism map with different types of models used in different regions.

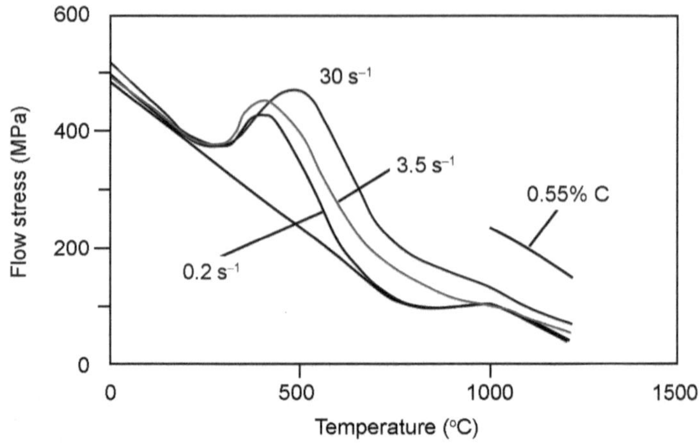

9.14 Flow stress of carbon steel with 0.15% C for different strain rates, from Suzuki *et al.* [299].

The yield stress usually decreases with temperature and the viscous effect becomes more important. However, anomalies may occur. One example where the yield stress becomes higher in some temperature regions can be seen in Fig. 9.14. The first peak at about 500 °C corresponds to dynamic strain ageing and the smaller peak around 1000 °C is due to the phase change. The dynamic strain ageing is also associated with decrease in flow stress with higher strain rates as well as oscillations in the measured stress–strain curve.

High-temperature data are sometimes hard to obtain. It may be possible to find similar material whose properties can be used. Figure 9.15 shows one

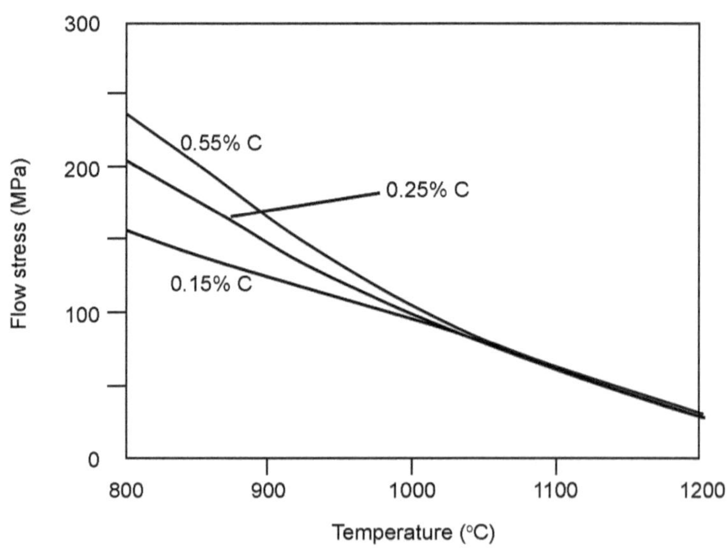

9.15 Flow stress of carbon steels, from Suzuki *et al.* [299].

example where the materials become more similar at higher temperatures. However, great care has to be taken. Sometimes, a small amount of an element may have been added to a variant of a material in order to give it extra high flow strength at higher temperatures. Then these materials will behave quite differently despite a similar chemical composition.

Most papers in CWM use the common rate-independent plasticity model based on von Mises yield criterion with the associated flow rule. The use of rate-dependent plasticity is less common, and some examples are given below. Argyris *et al.* [300, 301] used an overstress model of the form

$$\dot{\epsilon}^{\mathrm{p}} = \mu\left\langle \left(\frac{\bar{\sigma}}{\sigma^{\mathrm{y}}}\right)^2 - 1 \right\rangle \tag{9.8}$$

They used a Ramberg–Osgood relation for the evolution of the yield stress.

Chidiac and Mirza [302] used

$$\dot{\epsilon}^{\mathrm{p}} = AS_0\left\langle \frac{\bar{\sigma}}{\sigma^{\mathrm{y}}} - 1 \right\rangle^n \mathrm{e}^{-Q/RT} \tag{9.9}$$

where A is related to grain size, Q is the activation energy for grain growth and R is the gas constant.

Myhr *et al.* [303, 304] used a viscoplastic material model

$$\dot{\epsilon}^{\mathrm{p}} = \left(\frac{\bar{\sigma}}{A(\text{microstructure})F(\theta)(\bar{\epsilon}^{\mathrm{p}}_0 + \bar{\epsilon}^{\mathrm{p}})}\right)^{1/m(\theta)} \tag{9.10}$$

which is coupled to models for microstructure evolution affecting hardening. The strain hardening has a starting value $\bar{\epsilon}^{\mathrm{p}}_0$ and no increase in $\bar{\epsilon}^{\mathrm{p}}$ is allowed at temperature above a critical temperature.

Wang and Inoue [271] and Inoue and Wang [269] used a Perzyna type of model

$$\dot{\epsilon}^{\mathrm{p}} = \mu\langle\Phi\rangle \tag{9.11}$$

Goldak *et al.* [305, 306] discussed the use of different rate dependencies for the plastic behaviour at different temperatures and stresses. They chose to use different constitutive models for different temperature regions. A linear viscous model was used at a homologous temperature above 0.8. Rate-dependent plasticity was used down to a homologous temperature of 0.5 and von Mises plasticity for lower temperatures. Mahin *et al.* [289, 290], Winters and Mahin [307], Ortega et al. [199, 229] and Dike *et al.* [186] used a unified creep-plasticity model by Bammann and coworkers [308, 309]. Sheng and Chen [18] and Chen and Sheng [19, 310] used the Bodner–Partom viscoplastic model and the Walker model. The latter model accounts for kinematic hardening.

The material near and in the weld is subjected to reversed plastic yielding during the cooling phase. Thus using kinematic, isotropic or combined hardening will affect the stresses in this region. Bammann and Ortega [311] investigated the effect of assuming isotropic and kinematic hardening. They found that the choice of hardening influences the residual stresses a great deal in the weld metal, but further away the different models gave identical results. Devaux *et al.* [273] found only small differences in residual stresses between models with isotropic or kinematic hardening. They studied weld repair with four or six beads. Most studies use linear, isotropic hardening. Friedman [125, 126] assumed isotropic hardening using a power law hardening and Ueda *et al.* [263] used piecewise linear, isotropic hardening. Kinematic hardening [169, 170, 189, 258, 271] and combined hardening [187, 201, 312] has been assumed in some simulations. Vincent *et al.* [69] found that combined hardening gave a better agreement when using Satoh tests for calibration.

Further details about models and data are given in Lindgren [292]. However, it should be noted that the problem is not a shortage of models but the lack of parameters for the models.

9.7.2 Transformation-induced plasticity

Transformation-induced plasticity (TRIP) is loosely defined as 'a significantly increased plasticity during a phase change at a stress lower than normal yield stress' [313]. Early work used a heuristic approach by artificially lowering the yield limit but this is not a correct procedure. The TRIP phenomenon is due to local, microscopic stresses introduced to maintain the compatibility between the parent and the new phases. The TRIP effect is attributed to two parts: an accommodation effect, the Greenwood–Johnson mechanism [314], and an orientation effect, the Magee mechanism [315]. The volume difference between the phases generates plasticity in the weaker phase at low stresses in the Greenwood–Johnson mechanism. The Magee effect is caused by the formation of preferred orientations of the new phase that do not average to zero when a macroscopic stress is applied.

Oddy *et al.* [316] showed clearly the importance of including the TRIP effect in a 3D model of a butt-welded plate (see Fig. 9.16). This is confirmed by others later [177]. Sjöström included the TRIP effect in simulation of quenching in 1984 [317, 318]. It is believed that the Magee mechanism can be ignored in diffusional transformations but not for the formation of bainite and martensite [319]. Taleb and Petit [320] investigated this by checking whether the TRIP effect accumulated or not during thermal cycling with repeated austenisation. Their hypothesis was that the Magee effect would not accumulate, whereas the Greenwood–Johnson effect would. They found a strong accumulative effect supporting the hypothesis the Greenwood–Johnson effect dominated. Oddy *et al.* [321] discuss whether the shear may be self-accommodation and can be ignored even for the latter cases. The model by Leblond, described below, has been successfully used by many researchers.

It should be noted that all discussions below assume a two-phase composition with right subscript 1 denoting the parent phase and 2 the new phase. The papers

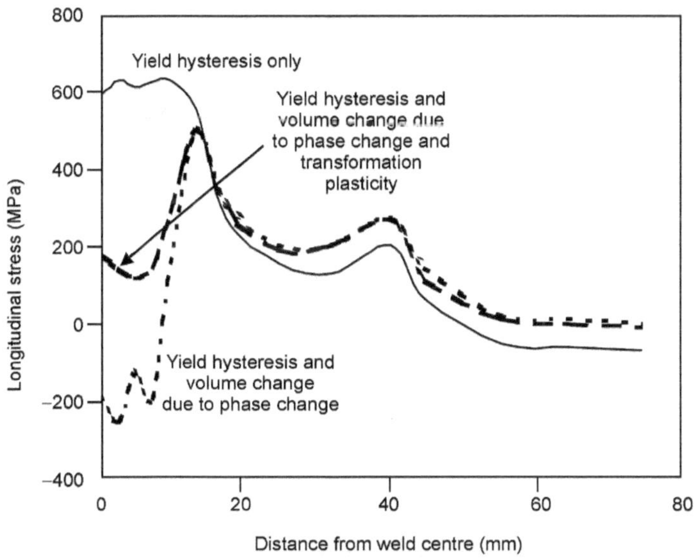

9.16 Residual longitudinal stress due to weld [316].

referenced below deal with a parent phase of austenite and martensite or bainite are the new phase. Naturally, the phase fractions sum to 100%:

$$X_1 + X_2 = 1 \qquad [9.12]$$

The TRIP effect is not unique notation for CWM. There is intensive research concerned with so-called TRIP steels (e.g. [285]). These are high-alloy austenitic steels that are metastable at room temperature and low-alloy steels with retained austenite that also are metastable. Martensite formation is then triggered by the deformation of the material. Modelling this phenomenon requires modelling of the mechanical effect on the phase change. Even more advanced models are required in the study of shape memory alloys (SMA) where reheating restores the shape of the material by removing the plastic straining due to a TRIP effect. Fischer *et al.* [313] used a thermodynamic approach to formulate a TRIP model. Gibbs free energy depends on the strain energy and additional hardening terms [248]. They added a function, denoted *h*, for the interaction of the parent and new phases. They also added a term for the chemical energy related to the transformation kinetics

$$\rho_0 \left[\phi_{\mathrm{ch,o}} - X_2 \Delta\phi_{\mathrm{ch}}(T)\right] \qquad [9.13]$$

where X_2 denotes the fraction of the new phase, ρ_0 is the original density, $\phi_{\mathrm{ch,o}}$ is the chemical energy of the parent phase and $\Delta\phi_{\mathrm{ch}}(T)$ is the difference in chemical energy of the parent and the new phases. Furthermore, they introduced extensions of the potential for the yield surface and a similar potential for the transformation kinetics.

Solving an extremum problem that maximises the energy dissipation according to Clausius-Duhem's inequality together with the consistency condition for the two potentials gives constitutive equations for the internal variables. They [248] illustrated the approach by introducing specific forms of potentials for the yield surface and the transformation kinetics, leading to an expression for the transformation plasticity strain components

$$\dot{\epsilon}_{ij}^{\mathrm{tp}} = -\frac{F_{\mathrm{g}}}{G_{\mathrm{f}}}\left[s_{ij} - \alpha_{ij}\right]\dot{X}_2 + \mathrm{e}_{ij}^{\mathrm{tr\_rec}}\dot{X}_2 \tag{9.14}$$

where equations for F_{g} and G_{f} are given in [313]. The term $e_{ij}^{\mathrm{tr\_rec}}$ is the deviatoric part of the recoverable transformation strain. The backstress tensor comes from

$$\alpha_{ij} = \left(\frac{\partial H}{\partial \epsilon_{ij}^{\mathrm{p}}}\right)^{\mathrm{dev}} \tag{9.15}$$

where it is the contribution to Gibbs free energy due to the internal stress state.

The model corresponds to the model proposed by Leblond, described below, if the backstress and the recoverable transformation strains are set to zero. The last term accounts for the Magee effect and the backstress term can give a transformation plasticity strain rate even after unloading. That would explain the observations by Taleb and Petit [320] and others [322]. Taleb and Petit found that the model by Leblond (below) was unable to explain this. The chosen potential for the transformation potential leads to

$$\dot{X}_2 = k_0(1 - X_2)\left[\epsilon_{ij}^{\mathrm{tr\_rec}}\dot{P}_{ij} + \rho_0 \frac{\partial \Delta\phi_{\mathrm{ch}}}{\partial T}\dot{T}\right] \tag{9.16}$$

where P_{ij} is the second Piola–Kirchhoff stress tensor. Integration with initial conditions of zero stress and transformation start at M_{s} gives an extended Koistinen–Marburger equation [323] that includes the effect of stress on the martensite formation

$$X_2 = \left(1 - \exp\left[-k_0\epsilon_{ij}^{\mathrm{tr\_rec}}P_{ij} + k_0\rho_0\left|\frac{\partial \Delta\phi_{\mathrm{ch}}}{\partial T}\right|(T - M_{\mathrm{s}})\right]\right) \tag{9.17}$$

The transformation starts when

$$\epsilon_{ij}^{\mathrm{tr\_rec}}P_{ij} - \rho_0\left|\frac{\partial \Delta\phi_{\mathrm{ch}}}{\partial T}\right|(T - M_{\mathrm{s}}) = 0 \tag{9.18}$$

and the argument of the exponential equation in Eq. [9.17] is restricted to positive values.

The model developed by Leblond [286, 319, 324–327] is by far the most commonly used one in CWM. He rigorously derived the model with a homogenisation procedure and based on classical plasticity without postulating its existence. This led to a formulation that had the same properties as earlier heuristic formulas. The model only accounts for the Greenwood–Johnson effect.

It is described below for the case of isotropic hardening plasticity. Note that these equations are applied only when $\bar{\sigma} < \sigma^{y}$. The transformation-induced plasticity is included in the classical plasticity when the effective stress is equal to the macroscopic flow stress. The original observation in the uniaxial case, discussed in Leblond *et al.* [325], was formulated as

$$\epsilon_{xx}^{tp} = K\Psi(X_2)\sigma_{xx} \qquad [9.19]$$

with different proposals for K and the function Ψ. However, incremental relations for multiaxial cases are needed and above is generalised to

$$\dot{\epsilon}_{ij}^{tp} = \frac{3}{2}K\frac{d\Psi}{dX_2}s_{ij} \qquad [9.20]$$

The relation proposed by Leblond *et al.* [325] that will be used below is

$$\Psi = X_2(1 - \ln X_2) \qquad [9.21]$$

They derived an incremental relation which, expressed in the increment in the effective value of the transformation plasticity in the austenitic phase, became

$$\dot{\bar{\epsilon}}_1^{tp} = -\frac{1}{X_1}\frac{2\epsilon_{1\to 2}^{trv}}{3\sigma_1^{y}(\bar{\epsilon}_1^{p})}h\left(\frac{\bar{\sigma}}{\sigma^{y}}\right)\ln(X_2)\dot{X}_2\bar{\sigma} \qquad [9.22]$$

where $\epsilon_{1\to 2}^{trv}$ is the transformation volume change, i.e. the difference in specific volume between the parent and new phases, σ_1^{y} and $\bar{\epsilon}_1^{p}$ denote the yield limit and the effective plastic strain in the weaker (X_1) phase respectively. The function is not well defined at $X_2 = 0$ and the transformation rate is set to zero if $X_2 \leq 0.03$. The 0.03 limit is also motivated from approximations in the model [325]. The stress dependency function h is

$$h\left(\frac{\bar{\sigma}}{\sigma^{y}}\right) = \begin{cases} 1 & \text{if } \dfrac{\bar{\sigma}}{\sigma^{y}} \leq 0.5 \\ 1 + 3.5\left(\dfrac{\bar{\sigma}}{\sigma^{y}} - \dfrac{1}{2}\right) & \text{if } \dfrac{\bar{\sigma}}{\sigma^{y}} > 0.5 \end{cases} \qquad [9.23]$$

Another variant was proposed by Devaux [328] (cited by Taleb *et al.* in [223]). However, Taleb *et al.* did not find that it made a better fit with their data.

There are also additional terms for classical plasticity in the weaker parent phase. They are computed from

$$\dot{\bar{\epsilon}}_1^{p\sigma} = \frac{1}{X_2}\frac{\bar{\sigma}}{\sigma_1^{y}(\bar{\epsilon}_1^{p})}\frac{\dot{\bar{\sigma}}}{E} \qquad [9.24]$$

$$\dot{\bar{\epsilon}}_1^{pT} = 2\frac{X_2\ln(X_2)}{X_1}\frac{\bar{\sigma}}{\sigma_1^{y}(\bar{\epsilon}_1^{p})}(\alpha_1 - \alpha_2)\dot{T} \qquad [9.25]$$

where $(\alpha_1 - \alpha_2)$ is the difference between the thermal expansion coefficients of the phases. This term can be ignored [325] as it is 25 times smaller than the transformation plasticity term. The effective plastic strain increment is used to compute the hardening in the austenitic phase. It is the sum of the above contributions and its contribution to the macro-effective plastic strain rate is

$$\dot{\bar{\epsilon}}^{\mathrm{p}} = X_1 \dot{\bar{\epsilon}}_1^{\mathrm{p}} = X_1 \left(\dot{\bar{\epsilon}}_1^{\mathrm{tp}} + \dot{\bar{\epsilon}}_1^{\mathrm{p}\sigma} + \dot{\bar{\epsilon}}_1^{\mathrm{pT}} \right) \qquad [9.26]$$

Computing component values from effective values is done, for example, by

$$\dot{\epsilon}_{ij}^{\mathrm{p}} = \frac{3}{2} \dot{\bar{\epsilon}}^{\mathrm{p}} \frac{s_{ij}}{\bar{\sigma}}, \; \dot{\epsilon}_{ij}^{\mathrm{tp}} = \frac{3}{2} \dot{\bar{\epsilon}}^{\mathrm{tp}} \frac{s_{ij}}{\bar{\sigma}}, \text{ etc.} \qquad [9.27]$$

The model ignores plasticity in the martensite phase when $\bar{\sigma} < \sigma_{\mathrm{y}}$. All contributions above are included in the classical plasticity term during global plastic yielding, $(\bar{\sigma} = \sigma_{\mathrm{y}})$. Thus these equations are not used then (see also Leblond and coworkers [327, 329, 330]). It should, however, be noted that the use of a linear mixture rule for phase properties is assumed above. The evolution of the yield stress in martensite is also computed. Thus the questions about choice of mixture rule and inheritance of structure from parent phase are important and the discussion in Section 9.3 is relevant. This is discussed in Leblond [326] for the case $\bar{\sigma} = \sigma_{\mathrm{y}}$.

Figures 4.9, 4.10 and 8.6 show the stress during a heating/cooling cycle for the Satoh test when including the transformation plasticity. Its effect is to reduce the magnitude of the stress as it gives a plastic strain component in the deviatoric stress direction. The logic used when computing the stresses in Fig. 8.6 is based on Leblond's model and the implementation in a finite element context can be seen in routine LEL_TransfPlast in CWM_Lab. Vincent *et al.* [319] shows how to use the model when using a viscoplastic material model.

An extended variant in order to include a transformation shear effect in the accommodation term [313, 331, 332]:

$$K = \frac{5}{6} \frac{\sqrt{\left(\epsilon_{1\to 2}^{\mathrm{trv}}\right)^2 + 3/4 \left(\gamma_{1\to 2}^{\mathrm{tr}}\right)^2}}{\sigma_*^{\mathrm{y}}} \qquad [9.28]$$

$$\sigma_*^{\mathrm{y}} = \sigma_2^{\mathrm{y}} \left(\frac{1 - \sigma_1^{\mathrm{y}}/\sigma_2^{\mathrm{y}}}{\ln\left(\sigma_1^{\mathrm{y}}/\sigma_2^{\mathrm{y}}\right)} \right) \qquad [9.29]$$

where $\gamma_{1\to 2}^{\mathrm{tr}}$ is transformation shear strain, σ_1^{y} and σ_2^{y} are the yield limit of austenite and martensite phases.

Otsuka *et al.* [225] generalises to a mixture of N phases, each with its own contribution to the transformation plasticity. They are ordered sequentially, $I = 1, 2 \ldots N$, into phases from lower to higher yield strength and each one with a yield function

$$f_I \left(\epsilon_{I_{ij}}^{\mathrm{tp}}, \bar{\epsilon}_I^{\mathrm{tp}}, \sigma_{ij}, T, X_{I+1}, X_{I+2}, \ldots X_N \right) \text{ for each } I = 1 \ldots N \qquad [9.30]$$

This is combined with the consistency condition during plastic yielding, $f_I = 0$, and the flow rule for associative plasticity. This leads to a plastic strain rate in each phase

$$\dot{\epsilon}_{ij}^{\mathrm{tp}} = \sum_{J=I+1}^{n} \frac{3}{2} K_J \dot{X}_J s_{ij} \qquad [9.31]$$

Summation over all phases gives

$$\dot{\epsilon}_{ij}^{\mathrm{tp}} = \sum_{I=2}^{n} X_I \dot{\epsilon}_{I_{ij}} = \sum_{I=1}^{n} \frac{3}{2} K_I \left(\sum_{J=1}^{I-1} X_J \dot{X}_J \right) s_{ij}$$

And in the case of two phases, $N = 2$, they get

$$\dot{\epsilon}_{ij}^{\mathrm{tp}} = \frac{3}{2} K X_1 \dot{X}_1 s_{ij} = -\frac{3}{2} K (1 - X_2) \dot{X}_2 s_{ij} \qquad [9.32]$$

9.8 Thermomechanical properties

Thermal dilatation is the coupling from the thermal problem to the mechanical part. It consists of a thermal expansion together with volume changes due to phase changes. This is utilised in free dilatation measurements to identify phase changes [61] (see also Section 4.1). Note that the thermal dilatation curve is the uniaxial measured thermal dilatation which is one-third of the volume change. Thus the difference in specific volume, $\epsilon_{1\to 2}^{\mathrm{trv}}$, used in the TRIP model in previous section is three times larger than measured in a uniaxial case. The thermal expansion coefficient is the derivative of the thermal dilatation curve with respect to temperature. This is sometimes called the tangent thermal expansion coefficient in contrast to the secant coefficient. The latter is the slope of a line going from the point at the thermal dilatation curve at a given temperature to the reference temperature. Care should be taken to control what kind of information the used finite element code needs: thermal dilatation curve or tangent or secant thermal expansion coefficient. The first option is preferable as this is easiest to compute from dilatation measurements. An illustration for the values for the thermal dilatation for different steels is given in Fig. 9.17. The wide field of variations for mild and low-alloy steels at higher temperatures is due to their variations in phase changes that are sensitive to the alloying of the steel.

9.9 Microstructure evolution

Coupled thermo-metallurgical-mechanical simulations (see Section 3.3), need models for the microstructure evolution. The kind of microstructure models that can be used in large-scale simulations are density types. An exception is Thiessen *et al.* [333]. They combined phase field models of the microstructure evolution. The phase field domains were located at the nodes of the finite

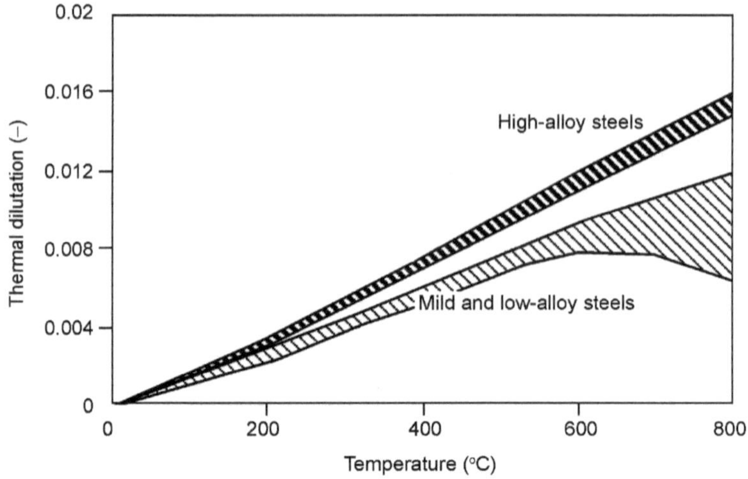

9.17 Thermal dilatation of steels during heating [254].

element model and thus did not cover the whole discretised volume. They did not apply a coupled TMM model as the microstructural information was not used to compute thermal and mechanical properties.

The focus below is on the solid state transformation of carbon steels. These models include only grain growth, precipitates and carbon diffusion when the material is fully austenised. The melting/solidification is naturally also an important phase change where a lot has been done regarding modelling of casting processes. This is discussed briefly in Section 9.10. The grain growth and precipitate models are referenced in Section 9.9.3.

The microstructure is represented by phase fractions, X, grain sizes g, precipitate size, d, etc. in the density type of models. These are density fields, usually computed at the integration points of the elements. Their values are computed at each point separately, i.e. no coupled system of equations need be solved. However, it should be noted that the differential equations used are often so-called stiff equations. For example, the transformation rate of a phase can vary by several orders of magnitudes. Therefore, robust and quite accurate methods for their integration are needed. 'Quite accurate' means that it is possible to waste computer time by reducing the error in the numerical integration algorithm in a way that is out of proportion to the precision of the data used to calibrate the model. Microstructure data are often in the form of time–temperature transformation (TTT), continuous cooling transformation (CCT), continuous heating transformation (CHT) diagrams. The TTT diagrams are also called isothermal transformation (IT) diagrams. An easily accessible information source for non-experts in metallurgy can be found in [334]. Usually the CCT diagrams are based on a soaking temperature around 900 °C appropriate for quenching. CCT diagrams for welding exist [59]. They are based on a higher hold temperature before the cooling. This temperature is usually between 1350 and 1400 °C in order to promote grain growth.

The two phase composition is taken as the starting approach in the descriptions below. Then one single phase transforms completely to another phase. The basic relation is

$$X_1 + X_2 = 1 \qquad [9.33]$$

The parent phase is denoted as phase 1 and the product phase as phase 2.

9.9.1 Model by Leblond

Leblond and coworkers [335, 336] discussed approaches that are based on Scheil's additivity rule. The transformation for a phase, X_2, is complete when

$$\int_0^1 \frac{d\zeta}{\tau[X_2, T(\zeta)]} = 1 \qquad [9.34]$$

where $\tau(X_2, T)$ is the time required to reach the fraction X_2 at a constant temperature T. The model cannot be used in the case of incomplete transformations. One simple example, from Leblond and coworkers [335, 336], is to assume a constant temperature of T_1 during $[0, \tau(X_2, T_1)/2]$ and then the constant temperature $T_2 > T_1$ during $[\tau(X_2, T_1)/2, \tau(X_2, R_1)]$. It is also assumed that $\tau(X_2, T_2) \to \infty$, i.e. the transformation will not be complete at this temperature. Then this model gives

$$\frac{\tau/2}{\tau} + \frac{\tau/2}{\infty} = \frac{1}{2} \neq 1 \qquad [9.35]$$

Thus the additivity rule does not work without modifications for incomplete transformations. The same criticism was applied to isokinetics models [335, 336]. These models are written in a separable form

$$\dot{X}_2 = f(T)g(X_2) \qquad [9.36]$$

Leblond *et al.* proposed a model that can be used for incomplete transformation where the final composition, X_{eq}, is the equilibrium fraction reached after holding for a long time at temperature T. The transformation rate is written as

$$\dot{X}_2 = \frac{X_2^{eq}(T) - X_2}{\tau(T)} \qquad [9.37]$$

where τ is a time constant for the rate of the transformation at this temperature. This will in the case of isothermal transformations lead to

$$X_2 = X_2^{eq}\left(1 - e^{-t/r}\right) \qquad [9.38]$$

where $t = 0$ when the transformation starts.

9.9.2 JMAK model

The semi-empirical JMAK model (Johnson–Mehl [337], Avrami [338] and Kolmogorov [339]) for nucleation and growth of a phase is written as

$$X_2 = 1 - e^{-kt^n} \quad \text{or} \quad X_1 = e^{-kt^n} \tag{9.39}$$

The equation can also be derived from assuming that the change in Gibb's free energy is the driving force of the transformation and that it is proportional to the transformation rate [37, 340]. The additivity rule is used when solving this equation, e.g. in quenching simulations [341]. The problem of describing incomplete transformations, discussed above, can be circumvented by normalising the transformation [37, 342, 343] so that it is complete at the equilibrium fraction of the product phase, X_2^{eq}

$$X_2' = X_2^{eq} X_2 = X_2^{eq}\left(1 - e^{-kt^n}\right) \tag{9.40}$$

This assumption gives a relation that corresponds to Leblond's model in Eq. [9.38] with $n = 1$. The equation gives the fraction of the product phase that is created, X_2', assuming there was 100% of the parent phase available for the transformation. This is a fictitious fraction and the real fraction is $X_2' \cdot {}^0X_1$. Then the final relation becomes

$$X_2 = X_2^{eq}\, {}^0X_1\left(1 - e^{-kt^n}\right) \tag{9.41}$$

The modification with ${}^0X_1$ corresponds to assuming that this austenite fraction would behave as a 100% austenite fraction in terms of transformation rates. This equation need to be further developed when implementing in an incremental context as in a finite element code.

9.9.3 Kirkaldy model

Goldak and Akhlagi [51] described the models for the solid state transformations in hypoeutectoid steels developed by Kirkaldy and coworkers [344–346], Watt *et al.* [347] and Henwood *et al.* [348]. Oddy *et al.* developed the model further [349]. The latter implementation was used by Börjesson and Lindgren [65] in welding simulations. The model needs the chemical composition as input in order to compute the equilibrium temperatures. These can be adjusted further if a measured TTT diagram is available. A separate validation of the microstructure model was done by comparing computed microstructure evolution during cooling with CCT data [65].

The models by Oddy *et al.* [349] are described in detail below. The models for the diffusion-controlled decomposition of austenite are described first, followed by the instantaneous austenisation model. There is also a transient austenisation model (see [349]). The Koistinen–Marburger equation is used for the martensite formation, as described below. The set of models can handle

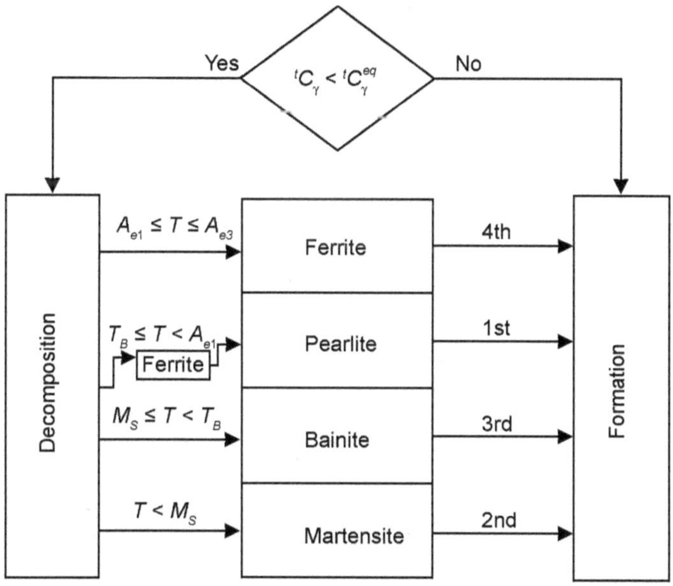

9.18 Austenite decomposition or formation.

partial austenisation and multiple reheating. They also include carbon diffusion as well as grain growth, since these influence the phase transformations strongly. There are two major parts in the algorithm; one is the logic for determining which reaction will take place (Fig. 9.18), and the other is the integration of the rate equations with appropriate expressions for the contributions to the transformation rates. It should be noted that the rate equations are stiff equations and the best to use is an adaptive integration algorithm. The evaluation of the phase fractions is done at the finite element nodes and/or integration points at every time step in the finite element simulation. These time steps are often too large for integration of these equations. An adaptive algorithm will split these time steps into smaller increments whenever needed [348].

The first step in the logic of the algorithm (upper part of Fig. 9.18) is to decide whether austenite will form or decompose. Austenite decomposes if its carbon content C_γ is lower than the equilibrium value C_γ^{eq} of the current temperature. The latter is given by the line G–S in the Fe–C equilibrium diagram in Fig. 9.19.

The decomposition of austenite is taken in the order indicated to the left in Fig. 9.18. Ferrite (denoted by subscript α) is created until the temperature reached A_{e3}, and thereafter pearlite forms (denoted by subscript P). The change in ferrite fraction is updated first, as indicated in the figure. The bainite reaction starts at the bainite transition temperature, T_B, which is the highest temperature at which the bainite appears in a TTT diagram. This temperature is defined [344] as the temperature at which the bainite transformation rate is larger than the ferrite and pearlite rates. The bainite reaction rate is zero at the bainite start

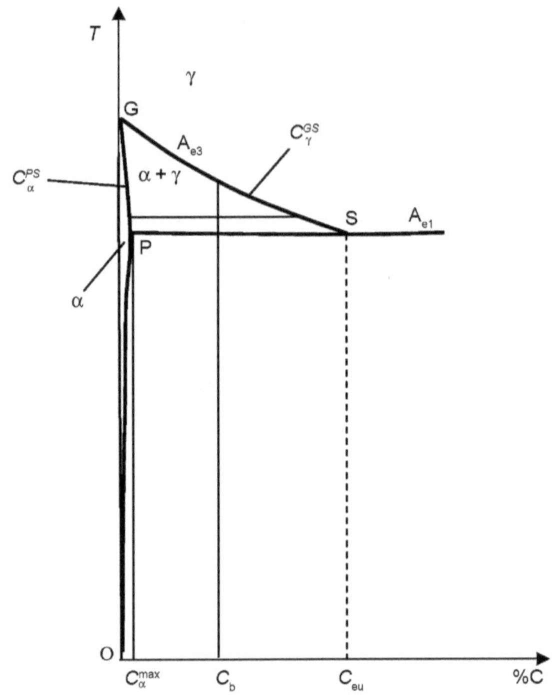

9.19 Fe–C equilibrium diagram for hypoeutectoid steel.

temperature B_s. The austenite which remains after reaching the martensite start temperature, M_s, produces martensite, denoted by subscript M, according to the Koistinen–Marburger equation. The critical temperatures A_{e1}, A_{e3}, B_s and M_s can be estimated from the alloy composition as [51, 347, 349, 350]

$$A_{e3} = 912 - 203\sqrt{C} - 15.2\mathrm{Ni} + 44.7\mathrm{Si} + 104\mathrm{V} + 31.5\mathrm{Mo} + 13.1\mathrm{W} - 30\mathrm{Mn} - 11\mathrm{Cr} - 20\mathrm{Cu} + 700\mathrm{P} + 400\mathrm{Al} + 120\mathrm{As} + 400\mathrm{Ti} \quad [9.42]$$

where, for example, C denotes the percentage of carbon. It is presumed in all equations in the following that the composition is given as a percentage:

$$A_{e1} = 723 - 10.7\mathrm{Mn} - 16.9\mathrm{Ni} + 29\mathrm{Si} + 16.9\mathrm{Cr} + 290\mathrm{As} + 6.4\mathrm{W} \quad [9.43]$$

$$B_s = 656 - 58\mathrm{C} - 35\mathrm{Mn} - 75\mathrm{Si} - 15\mathrm{Ni} - 34\mathrm{Cr} - 41\mathrm{Mo} \quad [9.44]$$

$$M_s = 561 - 474\mathrm{C} - 35\mathrm{Mn} - 17\mathrm{Ni} - 17\mathrm{Cr} - 21\mathrm{Mo} \quad [9.45]$$

However, it should be noted that other variants exist, e.g. [351], and it is best to have specific data for the material of interest.

The formation of austenite is computed according to the equilibrium Fe–C diagram in the instantaneous austenite formation model. This formation consumes the product phases in the order denoted to the right in Fig. 9.18.

The Kirkaldy model for the diffusion-controlled transformation of austenite decomposition into the product phases ($i = \alpha$, P, B, M) is written as

$$\dot{X}_i = F_i^G F_i^C F_i^T F_i^X \qquad [9.46]$$

where F_i^G is the effect of austenite grain size, F_i^C due to the chemical composition, F_i^T due to temperature, and F_i^X describe the effect of the current fraction formed.

The two-phase model is adapted to the multiple phase case where the transformation may finish at an equilibrium fraction of the product phases. The implementation by Oddy takes the austenite as the base variable; all transformations go either to or from austenite as shown in Fig. 9.18. The subscript γ denotes the austenite phase. Phase i corresponds to any of the product phases of ferrite, pearlite, bainite and martensite. Equation [9.33] is first modified by considering that some fraction of the material may have formed some phases that will not be participating in the current reaction. Thus the sum of the austenite and current product phase should be

$$X_\gamma + X_i = 1 - X_{\text{fix}} \qquad [9.47]$$

Furthermore, the parent phase may stop decomposing at an equilibrium value of X_γ^{eq} instead of disappearing completely. We write

$$X_\gamma - X_\gamma^{\text{eq}} + X_i = 1 - X_\gamma^{\text{eq}} - X_{\text{fix}} \qquad [9.48]$$

This can be rewritten as

$$\frac{X_\gamma - X_\gamma^{\text{eq}}}{1 - X_\gamma^{\text{eq}} - X_{\text{fix}}} + \frac{X_i}{1 - X_\gamma^{\text{eq}} - X_{\text{fix}}} = 1 \qquad [9.49]$$

Normalised variables according to above are used to give a relation in the standard form of Eq. [9.33]

$$\bar{X}_\gamma + \bar{X}_i = 1 \qquad [9.50]$$

This gives the relation between changes in parent and the product phase i as

$$\dot{\bar{X}}_\gamma = -\dot{\bar{X}}_i \qquad [9.51]$$

It is assumed that the rate equation Eq. [9.46] is applicable for the normalised variable, $\bar{X}_i$, to give the rate in Eq. [9.51]. Thereafter, this rate is re-scaled by $(1 - X_\gamma^{\text{eq}} - X_{\text{fix}})$ to give the real change in phase fractions. This can be expressed mathematically as

$$\Delta X_\gamma = \dot{X}_\gamma \Delta t = -\dot{X}_i \Delta t = -F_i^G F_i^C F_i^T F_i^{\bar{X}_\gamma} \cdot \left(1 - X_\gamma^{\text{eq}} - X_{\text{fix}}\right) \Delta t \qquad [9.52]$$

where $F_i^{\bar{X}_\gamma}$ denotes that this function is expressed in terms of normalised austenite fraction instead of a more commonly used fraction of the product phase. This replacement is based on Eq. [9.50]. The specific formulas used in

Table 9.4 Diffusion-controlled decomposition of austenite

	Ferrite	Pearlite	Bainite
$X_{fix} =$	${}^tX_P + {}^tX_B + {}^tX_M$	${}^tX_\alpha$	${}^tX_\alpha$
$X_\gamma^{eq} =$	${}^tX_\gamma \dfrac{{}^tC_\gamma - {}^tC_\alpha^{PS}}{{}^tC_\gamma^{GS} - {}^tC_\alpha^{PS}}$	${}^tX_\gamma \dfrac{C_{eu} - {}^tC_\gamma}{C_{eu} - {}^tC_\alpha^{PO}}$ [1]	0
$F_i^C =$	$\dfrac{1}{(59.6\text{Mn} + 1.45\text{Ni} + 67.7\text{Cr} + 244\text{Mo})}$	$\dfrac{1}{(1.79 + 5.42(\text{Mn} + \text{Mo} + 4\text{MoNi})f_P)}$ [2]	$\dfrac{1}{(2.34 + 10.1\text{C} + 3.8\text{Cr} + 19\text{Mo})f_B}$ [3]
$F_i^T =$	$(A_{e3} - T)^3 \exp[-Q_\alpha/RT]$	$(A_{e1} - T)^3 \exp[-Q_P/RT]$	$(B_s - T)^2 \exp[-Q_B/RT]$
$F_i^G =$	$2^{(G-1)/2}$ [4]	$2^{(G-1)/2}$ [4]	$2^{(G-1)/2}$ [4]
$F_i^{\bar{X}_\gamma} =$	$(1 - {}^t\bar{X}_y)^{(2{}^t\bar{X}_y)/3}\, {}^t\bar{X}_y^{[2(1-{}^t\bar{X}_y)]/3}$	$(1 - {}^t\bar{X}_y)^{(2{}^t\bar{X}_y)/3}\, {}^t\bar{X}_y^{[2(1-{}^t\bar{X}_y)]/3}$	$\dfrac{(1 - {}^t\bar{X}_y)^{(2{}^t\bar{X}_y)/3}\, {}^t\bar{X}_y^{[2(1-{}^t\bar{X}_y)]/3}}{\exp[C_r(1 - {}^t\bar{X}_y)^2]}$

[1] Oddy *et al.* [349] used the value C_α^{max} instead of ${}^tC_\alpha^{PO}$.
[2] $f_P = 1 + (0.01\text{C} + 0.52\text{Mo})e^{-Q_{mod}/RT}$.
[3] $f_B = 10^4 e^{X^2(1.9C+2.5Mn+0.9Ni+1.7Cr+4Mo-2.6)}$.
[4] G is the ASTM grain size number computed as $G = 1 + 1.4427 \cdot \log[shape \cdot (25.4/g)^2]$, where $shape = 100$ for square grains and $400/\pi$ for spherical grains; g is the grain size in microns.

Eq. [9.52] for each of the diffusion-controlled decompositions of austenite are given in Table 9.4.

The pearlite, bainite and martensite are assumed to have carbon compositions equal to C_{eu} (see Fig. 9.19) as they are formed from austenite with an eutectoid composition. The notations for carbon fraction in Table 9.4 are explained in Fig. 9.19. The variation with temperature of the ferrite carbon content was ignored by Oddy *et al.* [349]. They used $C_\alpha \equiv C_\alpha^{max}$, where C_α^{max} is the maximum carbon solubility in ferrite as shown in Fig. 9.19. The algorithm tracks changes in carbon content during the transformations.

$$^{t+\Delta t}C_\gamma = \frac{^tX_\gamma\,^tC_\gamma + \Delta X_\gamma C_i}{^{t+\Delta t}X_\gamma} \qquad [9.53]$$

The grain size of the austenite is also updated continuously. They used

$$\dot{g}_\gamma = \frac{k}{2g_\gamma}\,\mathrm{e}^{-Q_g/RT} \qquad [9.54]$$

from Easterling *et al* [352]. Grain growth among other metallurgical phenomena are described in Easterling [59], Svensson [353] and Evans and Bailey [354]. The equation is not applied until the temperature is higher than the solvus temperature for precipitates in the alloy. Models for dissolution, nucleation and growth of precipitates are described in Grøng [60].

9.9.4 Martensite formation

The martensite formation is primarily a thermal-driven transformation but it is also influenced by the stress state. It is a nucleation-controlled process, sometimes called a military transformation, and a martensite region is formed immediately once it is nucleated. The formation process is called *stress-assisted* nucleation when the nucleation is on the same sites responsible for a thermal-driven transformation. This is more important at lower temperatures nearer the M_s temperature, whereas the *strain-induced* martensite transformation is active at higher temperatures. The latter produces nucleation at new sites at the intersection of shear bands created by the plastic deformation [355] but mechanical twins can also be nucleation sites [356]. It is believed [357] that ϵ-martensite forms first at stacking faults of γ phase and α'-martensite is then formed at shear band intersections. It is also likely that ϵ-martensite transforms into α'-martensite at larger strains. This phenomenon is utilised in TRIP steels where metastable austenite forms martensite when plastically deformed. Strain-induced formation is not included in the description below, but inclusions of the effect of stress are shown.

The Koistinen–Marburger equation is nearly always the standard choice for the thermal-driven martensite formation [323]. Its standard format is

$$X_M = \{1 - \exp[-\alpha(M_s + M_s^\sigma - T)]\} \qquad [9.55]$$

where M_s is the start temperature of the martensite formation. The parameter α is typically set to 0.011. It is dependent on the carbon content of a steel and can be estimated from Figure 19.16 in Reed-Hill and Abbaschian [358] as

$$\alpha = \frac{4.61}{97.1C + 161} \qquad [9.56]$$

An extension of the model to include the effect of the stresses proposed by Fischer [359] and Tanaka and Sato [360] (see Eq. [9.18]), is to include a change in the martensite start temperature due to the stresses.

$$M_s^\sigma = \epsilon_{ij}^{\mathrm{tr\_rec}} \sigma_{ij} \qquad [9.57]$$

Another variant to include the effect of pressure and effective stress, used by Inoue and Wang [361, 362] and Denis *et al.* [363], is

$$M_s^\sigma = C_1 \sigma_m + C_2 \bar{\sigma} \qquad [9.58]$$

where C_1 and C_2 are material constants, σ_m is the mean stress and $\bar{\sigma}$ is the von Mises effective stress.

The rate form of Eq. [9.55] with M_s^σ according to Eq. [9.58] is [37]

$$\dot{X}_M = (1 - X_M)\alpha\left(C_1\dot{\sigma}_m + C_2\dot{\bar{\sigma}} - \dot{T}\right) \qquad [9.59]$$

The Koistinen–Marburger equation needs to be reformulated to rate-form in order to be useful for a more general temperature history. Oddy *et al.* [349] used the austenite fraction as the reference phase and extending their formulation with the M_s^σ term gives

$$X_\gamma = \exp[-\alpha(M_s + M_s^\sigma - T)] \qquad [9.60]$$

It is assumed that the fraction of the austenite existing when M_s is reached, denoted $X_\gamma^{M_s}$, follows the Koistinen–Marburger equation. An incremental relation can then be formulated as

$$\begin{aligned} \Delta X_\gamma &= X_\gamma^{M_s} \exp[-\alpha(M_s + M_s^\sigma - T - \Delta T)] \lfloor \exp[-\alpha(\Delta M_s^\sigma - \Delta T)] - 1 \rfloor \\ &= {}^t X_\gamma \lfloor \exp[-\alpha(\Delta M_s^\sigma - \Delta T)] - 1 \rfloor \end{aligned} \qquad [9.61]$$

Box 9.1 Computed TTT diagram using implementation by Oddy *et al.* [349].

The computation of a TTT diagram requires the evaluation of the rate equations given above and finding at what time a given percentage of a phase is reached. Phase boundaries can be 0.1% and 99.9% of the phases as shown in Fig. 9.20.

The diffusive transformations have to be evaluated by integration of Kirkaldy's equations. The martensite formation is only a function of temperature and these lines will be drawn as horizontal lines.

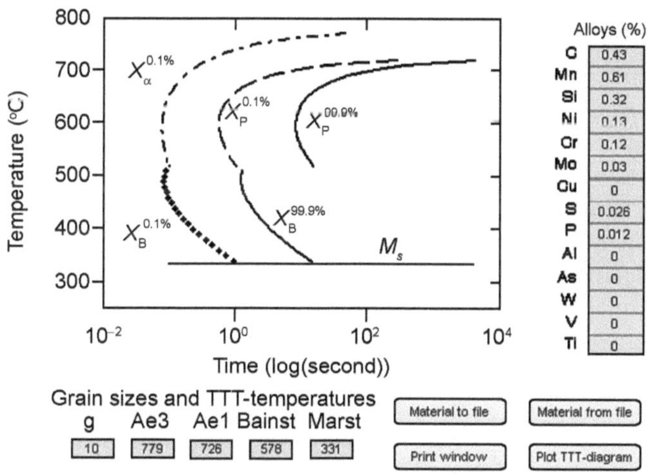

9.20 Computed TTT diagram.

We have for the first types of transformations the Kirkaldy model

$$\dot{X}_i = F_i^{\mathrm{G}} F_i^{\mathrm{C}} F_i^{\mathrm{T}} F_i^{\mathrm{X}} \tag{9.62}$$

The grain size is assumed to be constant at these temperatures and the temperature term is constant. However, the composition contribution is dependent on X for bainite. The specific equations are given in Table 9.4.

We get

$$\frac{1}{F_i^{\mathrm{X}} F_i^{\mathrm{C}}} \, \mathrm{d}X_i = F_i^{\mathrm{G}} F_i^{\mathrm{T}} \, \mathrm{d}t \tag{9.63}$$

giving

$$\int_0^{X_\%} \frac{1}{F_i^{\mathrm{X}} F_i^{\mathrm{C}}} \, \mathrm{d}X_i = F_i^{\mathrm{G}} F_i^{\mathrm{T}} t_\% \tag{9.64}$$

and finally

$$t_\% = \frac{1}{F_i^{\mathrm{G}} F_i^{\mathrm{T}}} \int_0^{X_\%} \frac{1}{F_i^{\mathrm{X}} F_i^{\mathrm{C}}} \, \mathrm{d}X_i \tag{9.65}$$

The integral must be evaluated numerically. It can be noted that it must not be repeated for each temperature as the integral with respect to X is the same. The diagram is created by choosing a number of temperatures where the time to obtain the given percentage gives, together with the temperature, a point in the diagram.

9.10 Material modelling in the weld pool

The thermal behaviour in the weld pool must take into account the latent heat of melting/solidification [364]. Alloys have a temperature range over which this occurs, $[T_{\mathrm{solidus}}, T_{\mathrm{liquidus}}]$. There is a large latent heat associated with this range. Feng [365], Feng *et al.* [366] and Dike *et al.* [367] used an advanced model for the release of latent heat during solidification of aluminium. Their model gave a higher release of latent heat in the upper part of the T_{solidus} to T_{liquidus} interval. Otherwise it is typically assumed that the latent heat is evenly distributed over this interval.

There is a modelling consideration of the melting behaviour that sometimes has to be taken into account in a thermal analysis. The heat in the weld pool is not only conducted but also convected owing to the fluid flow. Several researchers have imitated this stirrer effect by increasing the thermal conductivity at high temperatures, e.g. Andersson [88], Leung and Pick [170]. Michaleris and DeBiccari [189] used different conductivities in different directions at varying locations in the weld pool. It should be noted that this counteracts the use of distribution functions for the source (see Section 9.12). The introduction of a high conductivity in the weld pool is completely unnecessary if the heat input temperature is prescribed. The heat of fusion is the largest and most important in welding simulations. It is, for example, 277 kJ/kg for a 1.2% C-steel with T_{solidus} of 1387 °C and T_{liquidus} of 1481 °C.

The mechanical behaviour of the weld pool is in the classical CWM approach grossly simplified. The deformation behaviour is replaced by a 'soft' solid and the discussion in the next chapter deals with the different kinds of constitutive models used. The solidification shrinkage is often ignored in CWM models where the details near the weld are not important. It is included in models concerned with hot or solidification cracking; see some examples in Section 12.3.

9.11 Surface properties

The classical boundary conditions in heat conduction are the convective and radiative heat losses. However, there may sometimes be contact with fixtures or between different parts of the welded component on the welded structure over which heat transfers occur. There have also cases where underwater welding has been simulated.

The classical boundary condition giving surface heat flux is

$$q_{\mathrm{n}} = h(T - T_{\mathrm{ref}}) + \epsilon\sigma\left(T^4 - T_{\mathrm{ref}}^4\right) \qquad [9.66]$$

where h is the heat transfer coefficient, ϵ is the emissivity factor, and σ is Stefan-Boltzmann's constant. Figure 9.21 shows examples of emissivity factors for different surfaces. The equation can be rewritten in a format more convenient for finite element implementation:

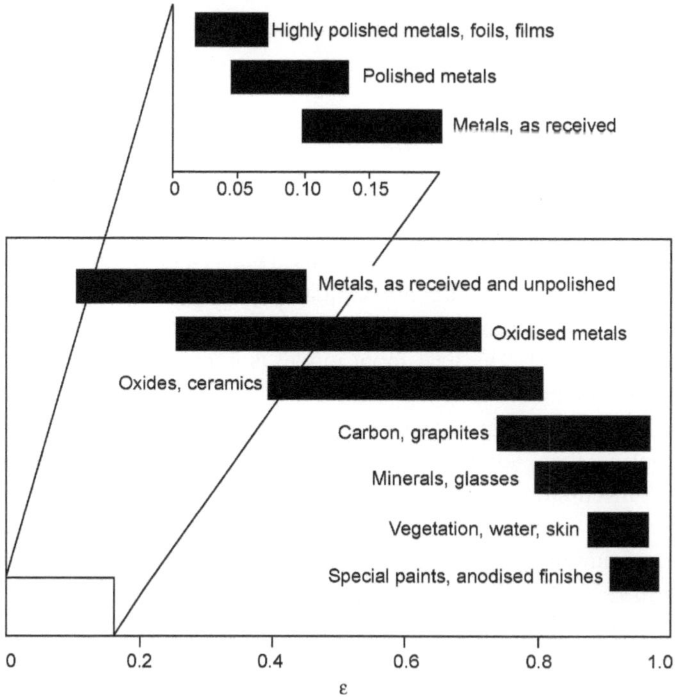

9.21 Emissivity coefficients of surfaces.

$$q_n = \left[h + \epsilon\sigma\left(T^2 + T_{\text{ref}}^2\right)(T + T_{\text{ref}})\right](T - T_{\text{ref}}) = h_{\text{eff}}(T)(T - T_{\text{ref}}) \qquad [9.67]$$

Textbooks about heat transfer gives a variety of combination for surfaces with moving fluids, contacts, etc. There are a few publications that include the effect of pressure on contact heat transfer [368–371]. The cooling due to water varies greatly depending on whether a vapour cloud is trapped on the surface or not [372]. Information about cooling due to spraying can be found in Bamberger and Prinz [373].

9.12 Heat input models

The concept of a heat input model in CWM is not only related to the heat flux in the thermal part of the simulations. It is also concerned with the modelling of addition of filler material and the simplification of the thermal and mechanical material behaviour. The latter has been discussed earlier. The basic idea is the replacement of a complex physical process with a much simpler one (Figs 3.9 and 3.10). This brings in not only approximations but also the need for calibration procedures to determine the heat flux as discussed in Section 8.5.2. The major concern is the introduction of correct net heat input into the computational model. Heat flux models with different detailed distribution are available, and some guidelines for the accuracy needed in this respect are given in Chapter 10.

9.12.1 Prescribed heat flux

Analytic solutions are discussed only briefly and then the two finite element approaches, prescribing heat flux or temperature, are exemplified. For additional information see the reviews of Goldak *et al.* [166, 374–378] and Lindgren [379]. Goldak *et al.* [375] and Akhlagi and Goldak [5] also include some discussion of advanced weld pool models. Additional useful information can be found in the books by Radaj [6, 7]. The heat input model is not a purely predictive method [374], as the net heat input is unknown. The best approach is to measure temperatures, observe microstructure changes, size of weld puddle, etc. The heat input model is then adjusted until a good agreement with experiments is achieved. A discussion of this calibration of the net heat input is given in Section 8.5.2.

Analytic solutions such as Rosenthal's [52, 53] for temperature fields have been used in several studies (see Lindgren [86] for details and references). They simplify the heat source geometry and ignore the temperature dependency of the thermal properties. These solutions may work well for regions away from the weld although the choice of parameters used in them is not obvious. However, they are not appropriate nearer the weld. There is no reason to use the analytic solutions in welding simulations as the numerical simulation of the thermal field is quite straightforward except for reduced accuracy models (mentioned in Section 10.1), where quick and simple solutions are needed.

Hibbitt and Marcal [127], Paley and Hibbert [380] and Andersson [88] used surface prescribed energy as heat input and an impulse equation for the heat contributed by the addition of filler. Usually some kind of ramp is used with linear increasing heat input for the approaching arc, constant heat input when the elements are melted and linear decreasing heat input when the arc is leaving the element. Shim *et al.* [264] evaluated the effect of this ramp on heating rate and peak temperature and found it was not important. Nickell and Hibbitt [262] and Friedman [381] used a Gaussian distribution for the surface heat input. Goldak *et al.* [382] proposed a more elaborate model, the so-called double ellipsoid heat source (see Fig. 9.22). It was later extended to arbitrary distribution functions [377]. Sabapathy *et al.* [383] modified the model for the case of weaved welding.

The double ellipsoid heat source (the energy distribution on the top surface as shown in Box 9.2), has been used in many papers [170, 189, 384]. The heat input is defined separately over two regions conveniently expressed by a local coordinate system moving with the heat source (see Fig. 9.23). One region is in front of the arc centre, $z' > 0$ and the other is defined behind the arc. The volumetric heat flux, q, in front of the arc centre, is

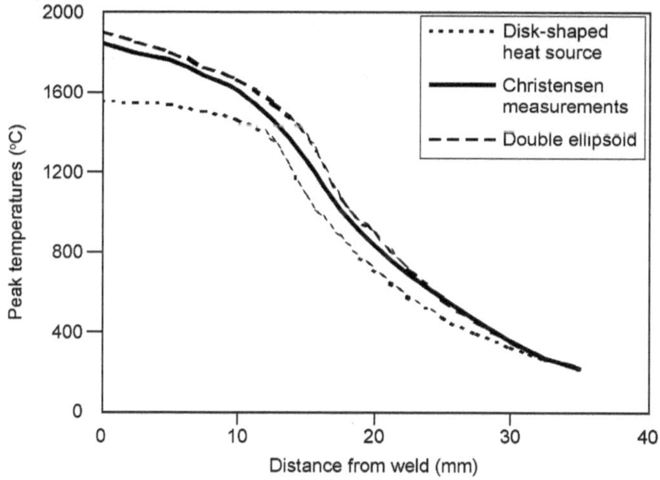

9.22 Comparison of heat input models and measured peak temperatures [382].

$$q(x', y', z') = f_1 \frac{6\sqrt{3}}{abc_1\pi^{3/2}} \dot{Q}_s \exp\left[-3(x'/a)^2\right] \exp\left[-3(y'/b)^2\right] \exp\left[-3(z'/c_1)^2\right] \quad [9.68]$$

and for the rear part

$$q(x', y', z') = (2 - f_2) \frac{6\sqrt{3}}{abc_2\pi^{3/2}} \dot{Q}_s \exp\left[-3(x'/a)^2\right] \exp\left[-3(y'/b)^2\right] \exp\left[-3(z'/c_2)^2\right] \quad [9.69]$$

where $\dot{Q}_s$ is the power of the source. Its relation to the net power of the finite element model is discussed below. The ellipsoid axes a, b, c_1 and c_2 are characteristic sizes of the weld pool. No heat flux is applied outside the volume of the double ellipsoid. The value of q at this cut-off boundary is 5% of the

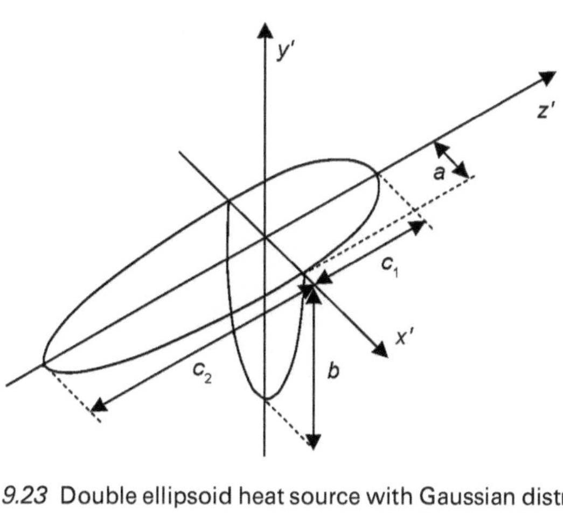

9.23 Double ellipsoid heat source with Gaussian distributed heat [382].

maximum q. f_1 and f_2 distribute the heat to the regions in front of and behind the arc centre. If a continuous heat flux along $z' = 0$ is wanted, then

$$\frac{(2 - f_2)}{c_2} = \frac{f_1}{c_1} \qquad [9.70]$$

Furthermore, if the total heat should be $\dot{Q}_s$ when integrating over infinity, then

$$f_1 + f_2 = 2 \qquad [9.71]$$

Combining above reduces the number of independent parameter of the model to the geometric measures and the value of $\dot{Q}_s$ as we get

$$f_1 = \frac{2}{1 + c_2/c_1} \qquad [9.72]$$

and

$$f_2 = 2 - f_1 \qquad [9.73]$$

The case of an arc moving along the z-axis with a speed of v gives a simple relation between the local and global coordinate system:

$$x' = x, y' = y, z' = z - vt \qquad [9.74]$$

Lundbäck [385] developed an algorithm for heat input based on discretisation of the weld path into line segments. Normal and tangential vectors are associated with the segments. A mapping between the global coordinate system and the local system could then be performed, facilitating the calculation of nodal heat input into the finite element model. Integration procedures are also needed for computing consistent nodal forces used in the finite element analysis. One example can be seen in Fig. 9.24, where the lighter areas are warmer owing to a moving heat source.

It is often necessary to take small time steps. Then a straightforward search for elements near the current arc location is sufficient. However, sometimes longer time steps can be taken or the welding velocity is artificially increased in order to

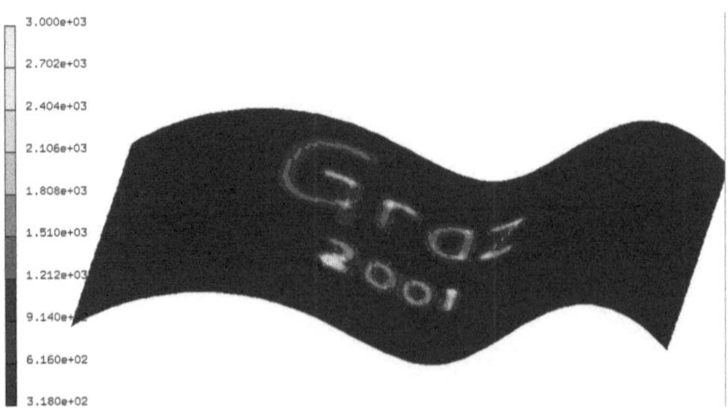

9.24 An illustration of the flexibility to prescribed path of heat source to a finite element model by user routines to the finite element code Marc [385].

apply complete weld strings simultaneously. Then a more advanced search logic to find affected elements is needed as the arc may pass over several elements during a time step. The logic for calculation of the heat to be applied accounting for the motion of the arc during the increment is illustrated in Box 9.2.

Integrating Eq. [9.68] over the ellipsoid volume expressed in the local coordinate system gives

$$\dot{Q}_1^{\text{theory}} = f_1 \frac{6\sqrt{3}}{\pi^{3/2}} \dot{Q}_s \times \int_V \frac{1}{a} \exp[-3(x'/a)^2] \frac{1}{b} \exp[-3(y'/b)^2] \frac{1}{c_1} \exp[-3(z'/c_1)^2] \mathrm{d}x' \mathrm{d}y' \mathrm{d}z \quad [9.75]$$

Introducing normalised coordinates gives a spherical volume of the source:

$$\begin{aligned}\dot{Q}_1^{\text{theory}} &= f_1 \frac{6\sqrt{3}}{\pi^{3/2}} \dot{Q}_s \int_V \frac{1}{abc_1} \exp[-3(x''^2 + y''^2 + z''^2)] a\mathrm{d}x'' b\mathrm{d}y'' c_1 \mathrm{d}z'' \\ &= f_1 \frac{6\sqrt{3}}{\pi^{3/2}} \dot{Q}_s \int_V \exp[-3(r'')^2] \mathrm{d}V''\end{aligned} \quad [9.76]$$

The integrand does not depend on the angle but only on the radius and the space angle of the half sphere is 2π. This leads to

$$\begin{aligned}\dot{Q}_1^{\text{theory}} &= f_1 \frac{6\sqrt{3}}{\pi^{3/2}} \dot{Q}_s \, 2\pi \int_0^1 \exp[-3(r'')^2] r''^2 \mathrm{d}\Psi \\ &= \frac{f_1 \dot{Q}_s}{2} \left(\operatorname{erf}\left(\sqrt{3}\right) - \sqrt{\frac{12}{\pi}} \mathrm{e}^{-3} \right)\end{aligned} \quad [9.77]$$

or

$$\dot{Q}_1^{\text{theory}} \approx \frac{f_1 \dot{Q}_s}{2} (0.9857 - 0.0973) = \frac{f_1 \dot{Q}_s}{2} 0.8884 \quad [9.78]$$

The results show that even if a fine mesh is used, the heat input will be less than 90% of the nominal $\dot{Q}_s$. This is not a problem as there is always a need to monitor and adjust the actual heat input, $\dot{Q}_{\text{FEM}}$, into the finite element model. The accuracy of the integration of the distributed heat flux depends on the mesh and time stepping. The variations due to these discretisation effects can be removed by scaling $\dot{Q}_s$. This is possible only if the actual heat input into the model can be monitored. Then $\dot{Q}_s$ in Eqs [9.68] and [9.69] should be scaled by the factor s below

$$s = \frac{\dot{Q}_{\text{wanted}}}{\dot{Q}_{\text{FEM}}} \quad [9.79]$$

where integration over the elements affected by the heat input during the time increment gives

$$\dot{Q}_{\mathrm{FEM}} = \frac{\Delta Q_{\mathrm{FEM}}}{\Delta t} = \frac{1}{\Delta t}\int_{V^{(e)}}\int_{^{n}t}^{^{n+1}t} q\mathrm{d}t\mathrm{d}V^{(e)} \qquad [9.80]$$

where the symbol $V^{(e)}$ denotes that the evaluation of the integral is made by the numerical integration procedure for the elements in the finite element model that are affected by the heat source.

The double ellipsoid model is often used for welding processes such as tungsten inert gas (TIG) with a fairly large weld pool. Laser and electron beam welding processes have a more concentrated energy distribution. The formation of key holes [16, 228] and other phenomena makes this an even more complex process than other fusion welding processes. A conical heat source with uniform heat flux with constant heat flux may be equally appropriate in these cases as some Gaussian distribution. The source is shown in Fig. 9.25. Furthermore, it should be noted that there is a relation between using a distribution function for the heat and the finite element discretisation. This is discussed in Section 10.2.1.

The heat flux for a conical source with $\dot{Q}_{\mathrm{s}}$ uniformly distributed is

$$q = \frac{3\dot{Q}_{\mathrm{s}}}{\pi\left(r_0^2 + r_1^2 + r_0 r_1\right)h} \qquad [9.81]$$

where r_0 is the radius at the upper surface, r_1 is the radius at the bottom of the heat source and h is the plate thickness or penetration depth.

The net heat to be supplied via a heat source model should be

$$\dot{Q}_{\eta} = UI\eta \qquad [9.82]$$

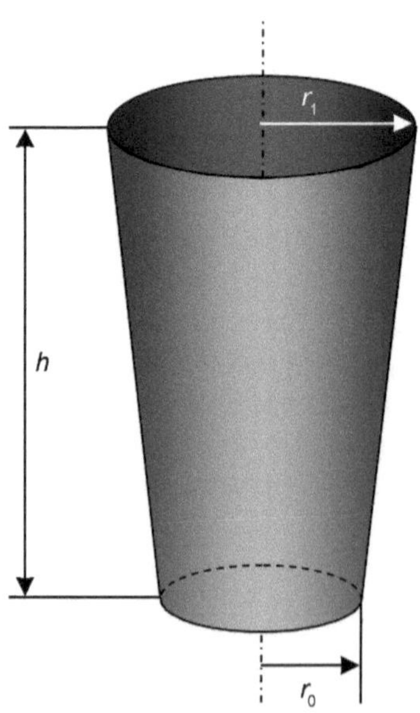

9.25 Conical heat source with constant heat input.

Table 9.5 Efficiency factors for different welding procedures from Grøng [60]

Welding process	SAW Steel	SMAW Steel	GMAW CO_2-steel	GMAW Ar-steel	GTAW Ar-steel	GTAW He-Al	GTAW Ar-Al
η	0.91–0.99	0.66–0.85	0.75–0.93	0.66–0.7	0.25–0.75	0.55–0.80	0.22–0.46

SAW, submerged arc welding; SMAW, shielded metal arc welding; GMAW, gas metal arc welding; GTAW, gas tungsten arc welding.

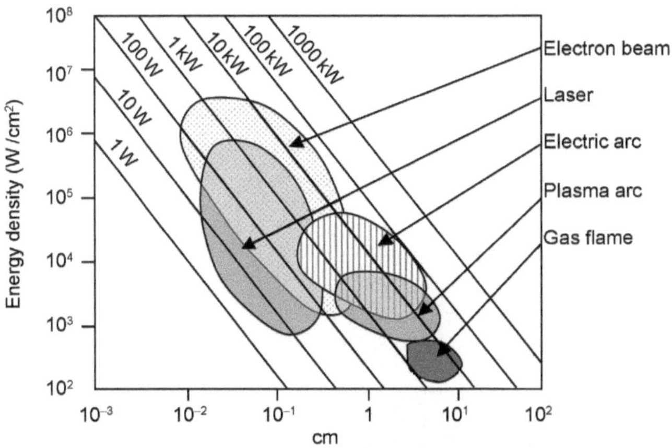

9.26 Energy densities and characteristic dimensions.

where U is the voltage, I is the current and η is an efficiency factor that depends on the local geometry and welding procedure. Some guidelines are given in Table 9.5 and the calibration procedures are discussed in Section 8.5.2. The differences between different welding processes are mainly expressed in different energy distributions (Fig. 9.26) in CWM models.

Box 9.2 Incremental logic for the double ellipsoid heat flux model

It is assumed that the heat source moves straight along a line during a time step. The local coordinate system of the heat source is used. The point with coordinates (x, y, z) in the global coordinate system moves then in the local system from (x', y', z'_1) to (x', y', z'_3) where $z'_3 = z'_1 - v\Delta t$. The energy input during the increment can be found by integration with respect to z' instead of time as

$$\int_{z'_1}^{z'_3} q(x', y', z') \frac{dz'}{v} \qquad [9.83]$$

where q is defined by Eqs [9.68] and [9.69]. The integration limits, z'_1 and z'_3 must be adjusted so that no heat input occurs when outside the double ellipsoid volume.

There is only one term in the integral Eq. [9.67] that depends on z'. The indefinite integral of that term is

$$\int e^{-3(z'/c_i)} \frac{dz'}{v} = \frac{1}{v} \frac{\sqrt{3}}{6} \sqrt{\pi} \cdot \operatorname{erf}\left(\frac{\sqrt{3}z'}{c_i}\right) \qquad [9.84]$$

where the subscript $i = 1$ for front part of source and 2 for rear part.

Two examples of this calculation of input energy for the parameters $v = 2.9$, $c_1 = 1$, $c_2 = 2$, $a = 2$, $b = 0.5$ and $\dot{Q}_s = 6$ are shown in Figs 9.27 and 9.28. The energy input during a time step is computed and its distribution over the top surface, $y \equiv 0$, is shown Figs 9.27 and 9.28. Figure 9.27 shows the results for a very small time increment, $\Delta t = 10^{-6}$. Figure 9.28 shows the energy distribution for $\Delta t = 2$. A very high velocity can be used to lay weld strings instantaneously, and with this approximation of the welding problem the computational effort reduces.

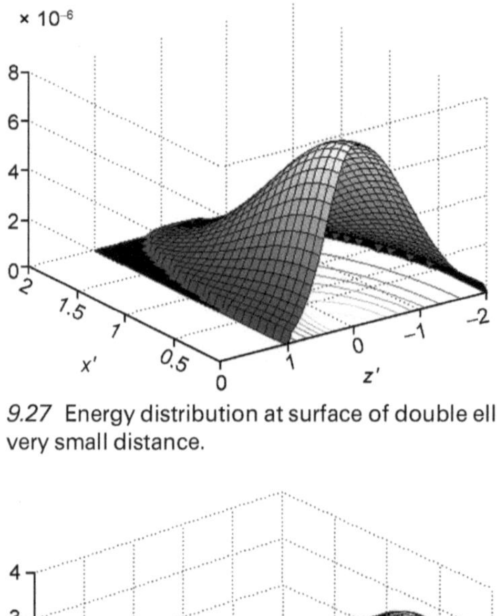

9.27 Energy distribution at surface of double ellipsoid heat source that has moved a very small distance.

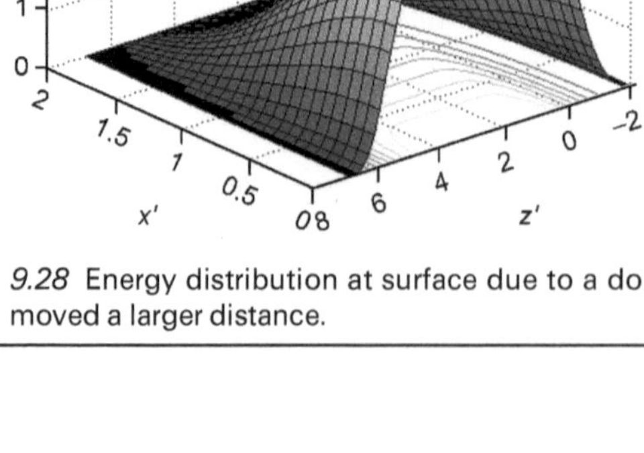

9.28 Energy distribution at surface due to a double ellipsoid heat source that has moved a larger distance.

9.12.2 Prescribed temperature

The notations in Fig. 9.29 are used to describe an approach for prescribed temperature as a means of heat input. The figure shows elements that are to be activated. They correspond to the cross-section of a weld where filler material is added. The same notations can be translated into a three-dimensional context. Then the volume of a set of elements correspond to a volume of filler material that is to be added during a time step. The cross-section of the current weld is denoted as Ω_w. This is a volume in the case of a three-dimensional model. The internal boundary between the weld and the existing material is denoted as Γ_{iw} and the free surface is denoted as Γ_{ew}. The intersection of the boundaries Γ_{ew} and Γ_{iw}, is denoted as Γ_{iew}. The latter is the two points A and B in the two-dimensional case and a space curve in the three-dimensional case. Γ_{iew} is defined as a subset of Γ_{iw} in the following.

The temperature is first raised at the set Γ_{iw} towards the weld from current temperature up to the prescribed temperature, T_{weld}. This is denoted by the linear increasing part of the line in upper diagram in Fig. 9.30. It starts at the time t_{start} and the heating of the boundary is achieved during a fraction of the total time for the prescribed temperature, denoted by ft_{weld}. The starting time t_{start} is the time when a weld is laid in a two-dimensional model and the total time for the prescribed temperature is estimated by

$$t_{weld} = \frac{l_{weld}}{v} \qquad [9.85]$$

where l_{weld} is the estimated length of the weld puddle and v is the welding speed. The fraction, f, for ramping up the surface temperature is a small value and chosen from numerical convenience. The interpretation of these times is different in three dimensional models. Then filler material is added every time step during the welding. The elements corresponding to the added filler material are thermally activated, and their nodes belonging to Ω_w are also given the

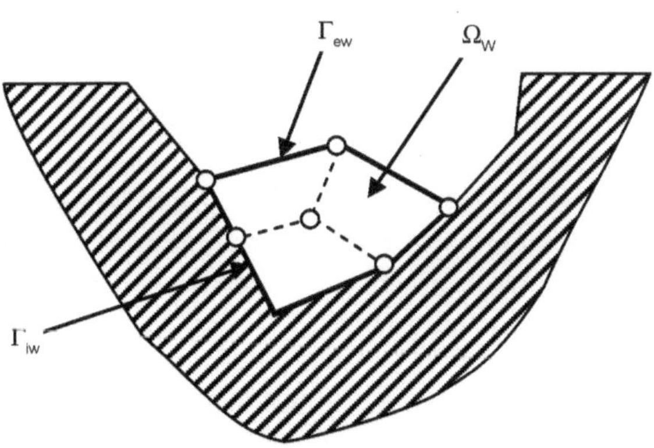

9.29 Notations used for definition of weld geometry.

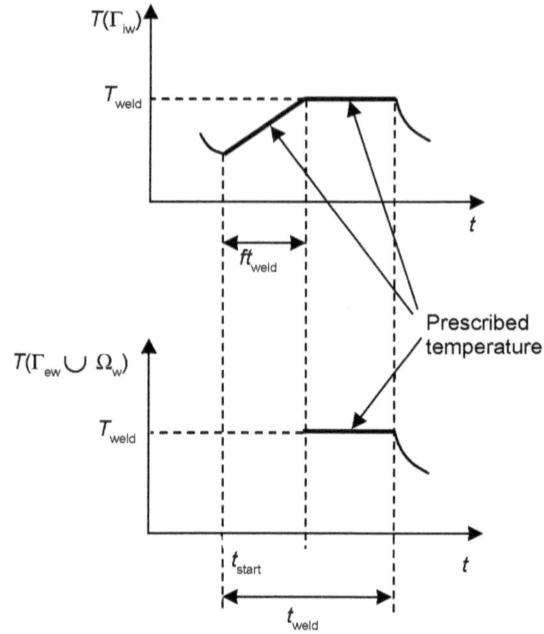

9.30 Prescribed temperature in weld [3]. Upper diagram is for nodes at boundary between weld and base metal. Lower diagram is for the rest of the weld. Notations are explained in Fig. 9.29.

prescribed temperature T_{weld}. This temperature is held for a given time and then the temperature is no longer prescribed but computed according to the heat conduction equation. Details about element activation are discussed in the next chapter.

Carmet *et al.* [386] and Goldak *et al.* [305] used prescribed temperatures. Jones *et al.* [387, 388] prescribed the temperature at the boundary of the weld pool to be equal to the $T_{liquidus}$. The boundary was defined beforehand. They simulated the welding of a bead on a disc. Temperatures were measured and compared with simulations. Quite good agreement with measured temperatures was obtained. Roelens and coworkers [174, 276, 277] and Lindgren *et al.* [3] prescribed the temperature in the case of multipass welds. All welds were assigned $T_{liquidus}$ 1520 °C in [3] when they were laid and the same length of the weld puddle was assumed for all weld passes even if the welding speed varied somewhat. Despite this, a reasonable good agreement with experiments was obtained (see Fig. 8.11).

Heat input models with a prescribed temperature cannot give the correct heating history in the weld region as this temperature increase is prescribed. However, this seldom has any influence on the subsequent history after melting.

9.12.3 Addition of filler material

The modelling of multipass welding poses some additional problems compared with single-pass welds. The material modelling due to the multiple thermal cycles causing repeated phase changes is one challenge in this respect. The other is the addition of filler material. All models of multipass welding are created in such a way that the whole model is defined initially. Thus elements corresponding to 'not laid' welds are included in this definition. They are removed from the analysis initially and then included automatically during the simulation. This is handled by deactivation and activation procedures. Two different approaches for imitating the addition of filler material can be used for this purpose: the *quiet* and the *inactive* element approaches. They were compared in Lindgren *et al.* [3] and were found to work equally well. They can be combined with prescribed heat flux or temperature for heat input as discussed above. The description below is based on the paper by Lindgren and Hedblom [148]. The filler material to be added at a given time step corresponds to the elements in Ω_w. This notation and others are explained in Section 9.12.2 and Fig. 9.29.

The so-called quiet element approach retains all elements and nodes in the problem definition. The elements corresponding to filler material that has not yet been laid are given properties at the start of the analysis so that they do not affect the surrounding structure and are assigned real material properties at the time the corresponding filler material is added. The displacement of the nodes in Ω_w and Γ_{ew} are computed at all times as they are always included among the unknown displacements. The so-called inactive element approach does not assemble these elements until the time the corresponding filler material is added. Then the profile of the equation system of the model must be recomputed.

It is preferable to activate the element in the thermal analysis before the mechanical. The need to supply heat to increase the temperature in the weld requires that the corresponding elements are active in the thermal analysis. Then it is best to delay their activation in the mechanical analysis until they have reached a high temperature such as $T_{solidus}$.

The quiet and inactive element approaches have the following pros and cons:

- An appropriate choice of material properties for elements corresponding to not laid welds has to be tried out in the quiet element approach. For example, the stiffness should be so small that this region does not exert any loading on the surrounding, but large enough that the resulting global stiffness matrix is well conditioned. This problem does not exist in the inactive element approach.
- The inactive element approach gives a smaller model. However, the recomputation of the matrix profile when adding filler material takes some time. This will be particularly cumbersome in the case of three-dimensional models where new material may be added in every time step.

Both methods require procedures that can relocate the nodes of these elements correctly although the inactive elements require larger adjustments in their positions. This is because all the nodes are given some displacements in the quiet element approach. The nodes corresponding to not laid filler material follow the deformation of the structure in some way, whereas they are in their initial location when using the inactive element approach. The method proposed by Lindgren and Hedblom [148] can be used for this purpose.

There is a need to ensure that the elements corresponding to added filler material have the correct location. It is not enough to define the element so that this is correct with respect to the initial geometry as this will not, in the case of a large deformation analysis, ensure the correct feed rate of filler material at the time when the weld is made. The transverse shrinkage of the joint can be considerable (see Fig. 9.31). Therefore it may even be difficult to define elements that have a reasonable shape in the initial groove so that they get the correct area/volume in the final weld seam. It may be possible to compensate for this by having an initial model defined, as in the left side of Fig. 9.31. But there are limits on how much these welds can be 'pressed down' and at the same time obtain the correct area/volume corresponding to the feed rate of filler material.

The basic idea (see [148] for details) is to use an optimisation procedure to relocate the nodes of the weld elements to be activated so that the wanted volume/cross-section of the weld is obtained while at the same time keeping the distortion of the elements low. Furthermore, constraints can be placed on the

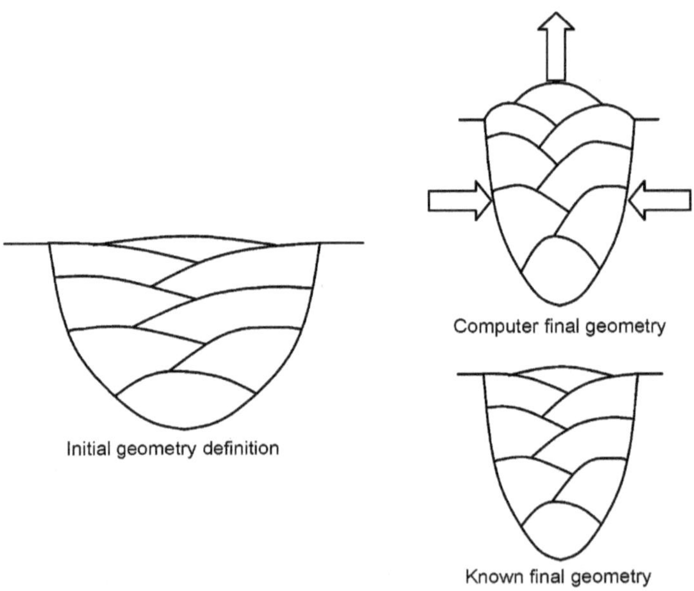

9.31 Effect of transverse shrinkage of groove and the problem of defining elements with appropriate area/volume in the initial geometry, from Lindgren and Hedblom [148].

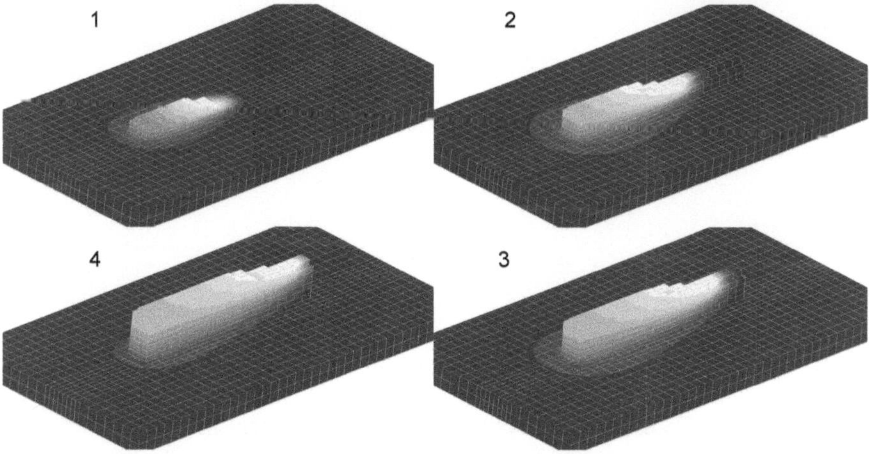

9.32 Example of element activation applied for a metal deposition process, courtesy A. Lundbäck. White denotes the highest, melting temperature, and black is room temperature.

wanted surface. Lindgren and Hedblom [148] included a requirement that the surface of the weld should be smooth. They minimised the function

$$w(\boldsymbol{p}) = D(\boldsymbol{p}) + f_1 \left[\frac{V_{\text{curr}}(\boldsymbol{p})}{V_{\text{wanted}}} \right] + f_2 \left[\frac{A_{\text{curr}}(\boldsymbol{p})}{V_{\text{wanted}}} \right] \qquad [9.86]$$

where $\boldsymbol{p}$ is the nodal coordinates of the nodes to be relocated, D is Oddy's distortion metric [152] given by Eq. [7.6], and f_1 and f_2 are weighting factors for the relative importance of obtaining the correct volume and smooth surface. It should be noted that the use of Oddy's distortion metric can cause problems if the elements are too distorted when the optimisation starts. This is particularly likely in the case of the inactive element approach.

The logic above can be used to simulate weld deposit processes as they are like multipass welding. One example of this is shown in Fig. 9.32. Several layers of welds are deposited with elements activated every time step. Excerpts from this history are shown, with the numbers showing their order. The greyscale denotes the temperatures.

9.13 Geometric models

The general solution of the nonlinear heat flow equations (Chapter, 5), and the equations for nonlinear deformations (Chapter 6), are based on the three dimensional versions of these equations. These correspond to the choice of solid elements in the finite element context. The driving force for reducing the dimension of the problem is to reduce the required computational effort. This is not reviewed here but is discussed in Sections 7.6 and 7.7.

10
Modelling strategy

Modelling in CWM poses several problems not seen in more traditional problems in mechanics. The first is that the stress response is the result of coupled thermal, metallurgical and deformation histories (Section 3.3 and Chapter 4). Welded structures are usually fabricated with many welds. These are either separate welds that intersect or are close enough to interact or are multipass situations where successive welds lie on top of one another. The multiple thermal cycles and sequence effects make analysis of real structures a challenge. The motion of the source also means that regions of very high gradients and rapid changes move through the structure. The goal of any intelligent modelling is to retain the essential features of the problem while at the same time removing non-essential or less significant facets until the cost of the analysis reaches tolerable levels. With each successive abstraction in the model hierarchy discussed in Section 8.3, there is always some reduction in the accuracy of the results. Therefore, there is a need for a strategy, a methodology, for the creation of models within the domain of classical CWM (Fig. 3.10).

This chapter aims to give guidelines for modelling in classical CWM. Dong *et al.* [190] stated that there is 'a great deal of confusion in the research community in terms of what level of detail is required to be modelled for the prediction of residual stresses and distortions'. This chapter will attempt to reduce this confusion. The accumulated experience of more than two decades shows that, despite some uncertainties, there is also a great deal of consensus in CWM.

The important concepts for modelling are defined in Section 8.1, of which the most important for this chapter is the scope of the model. This scope determines what kind of model and accuracy are required. The accuracy depends on the chosen model and values of the associated parameters. The values are related to the calibration process (Sections 8.5.1–8.5.3). Different modelling options are discussed in Chapter 9. There are several aspects of modelling of welding that must be considered (Fig. 10.1). The focus of the guidelines given in this chapter is on the most critical aspects; the heat input model, the material modelling and choice of geometry together with the discretisation issues. The first aspect is

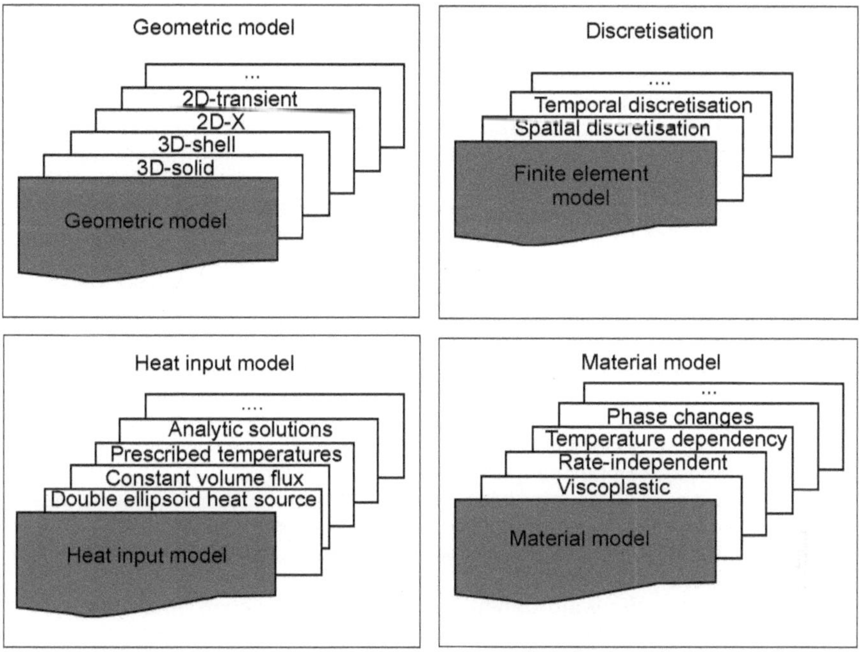

10.1 Different submodels of aspects important for the accuracy of CWM models.

specific to CWM models whereas the latter are common to all types of applications.

The proposed strategy consists of two strands, the determination of the scope of the simulation to be performed and the consideration of the weld flexibility of the component/structure to be welded. The first part is quite straightforward and leads to an accuracy classification for the model to be developed. Weld flexibility is an estimate of the effect of the heat source on the deformations. The proposed strategy is given in Section 10.1. Thereafter, the characteristics of these models are described followed by a discussion of the given guidelines.

10.1 Accuracy and weld flexibility categories for CWM models

The modelling strategy is summarised in the current section and some motivation for the proposed strategy is given in Section 10.3. The scope (see Section 8.1) of a model is a prerequisite. What question(s) should be answered by simulation? The scope and when in the design phase the model is used determine what kind of simplifications can be done. Questions that may be the motivation for performing a simulation of welding are given in Table 10.1. The questions are listed in order of increasing complexity. It may be found at this stage that a finite element

Table 10.1 Scope of a welding simulation

1. What are the residual stresses? These will be used for determination of risk for buckling, fatigue, cold cracking or stress corrosion cracking, etc.
2. What are the final deformations? These are wanted in order to check if the tolerance requirements on a component are fulfilled after the welding procedure.
3. What are the transient stresses and deformations? These may be wanted in order to find procedures that maintain the gap required for a successful weld.
4. What is the microstructure of the weld and the heat-affected zone? This kind of analysis also requires a microstructure model.
5. What causes hot cracking? Note that hot cracking phenomena are on the limit of continuum mechanics as they may be caused by liquid material along grain boundaries exposed to tensile strains. The parameters obtained from simulations must be evaluated with care in order to ascertain if the model can be used to study how the welding process can be changed to reduce hot cracking.
6. Is it possible to join these materials? The finite element method has no knowledge about metallurgy. It may be combined with other types of models that have this knowledge and use the computed results from finite element models.
7. What is the weld penetration? This is very much influenced by the chemical composition, addition of surface-active elements, arc physics, etc. It requires complex fluid mechanics models.

analysis cannot supply the wanted answers, because a thermomechanical model cannot answer the actual question(s) or that the necessary input, expertise or computational resource is lacking. The requirements on the model are also dependent on the effect of the temperature field on the deformations. This is captured by the geometry classification in Table 10.2. A structure is considered more flexible if the details of the temperature fields influence the deformation

Table 10.2 Weld flexibility categories

	Rigid	Standard	Flexible
Examples	Multipass welds[a]	One-pass welds	Loosely fixtured thin plates
Description	The cross-section of each string is small compared with the thickness of joined parts. The longitudinal residual stress in the weld reaches the yield limit[a].	Welds where the cross-section of the string is in the same size as the thickness of the joined parts.	Large deformations where variations in details of weld region will affect the deformations.

[a] The exception is the case of a low martensite start temperature. These multipass weld cases are classified as standard flexibility.

more. Section 4.2 discusses materials with low martensite formation temperatures. The basic assumption is that a three-dimensional model is needed, shell or solid element. Two-dimensional models can often be useful when evaluating strains and stresses and it is always a good practice to start with simple models in initial evaluations (see Section 7.6). However, in general, industrial applications require three-dimensional models using shell and/or solid elements.

The different accuracy categories are defined in Table 10.3. The choice of accuracy depends on the scope of the analysis, the questions in Table 10.1 and the kind of geometry analysed (Table 10.2). Reviewing the literature [86, 209, 292, 389] shows that there is some consensus on the requirements on models used for the three first points in Table 10.1. Therefore, there is general consensus in the literature about the characteristics of *basic* up to *accurate* simulations. The lower accuracy level, *reduced accuracy*, is associated with analyses at preliminary design stages. Although important, it is not discussed here. There is an overlap between the accuracy levels. The higher level should be chosen if details near the weld are needed. The *very accurate* level is not well established. Some papers concerned with this kind of model are reviewed in Section 12.3. The accuracy is mainly determined by the choice of heat input, material and geometric models together with the spatial and temporal discretisations.

The principal differences between basic models and standard accuracy models are that the latter have a finer mesh, smaller time steps and a more detailed material model and heat input model. The basic models completely ignore the 'high' temperature behaviour. The accurate and very accurate models are not only a refinement of the previous category but may also need to go in a higher level in the mesh hierarchy (Fig. 8.3) or include additional phenomena in the material model (Sections 9.2–9.10).

Table 10.3 Accuracy categories related to scopes and geometry categories

0. *Reduced accuracy levels.* Use of simple and fast models for evaluations at preliminary design stages. This part of the roadmap is not elaborated further in this book.
1. *Basic simulation* where only the overall residual state[a] is of interest, questions 1 and 2. This accuracy category is applicable for *rigid* components/structures.
2. *Standard simulation* is used when questions 1–3 in Table 10.1 are to be answered for rigid and *standard* components/structures.
3. *Accurate simulation* where the transient strains and stresses are wanted. This accuracy level is needed when questions 3–4 in Table 10.1 should be answered and also for questions 1 and 2 for flexible components/structures.
4. *Very accurate simulation* where the high temperature behaviour and the zone near the weld is important. This is necessary in order to answer questions 4 and 5.

[a] It may sometimes be necessary to have 3D models in order to obtain useful information about the deformations.

Evaluating the problem that should be modelled according to Tables 10.1 and 10.2 leads to an accuracy classification of the model as given in Table 10.3. The next section describes these categories with respect to the requirements of the heat input model, the material modelling and choice of geometry together with the discretisation issues.

10.2 Characteristics of different accuracy categories

The choices among the model options, reviewed in Chaper 9, for the different accuracy categories in Table 10.4 are summarised below.

10.2.1 Heat input model

Different heat input models are described in Section 9.12. Prescribed heat input models are straightforward to implement in any finite element code. Overall deformations and stresses are not particularly sensitive to the details in the heat distribution. The starting model is to use constant flux over a volume that corresponds to the weld pool. The correct amount of net heat input is the important issue. Section 8.5.2 discusses calibration of the heat souce model. Resolving a more detailed heat input distribution depends on the discretisation. The Gaussian double ellipsoid heat source model needs at least four quadratic elements along one semi-axis in order to resolve the Gaussian distribution of heat [166].

The proposal for heat input models are given in Table 10.4. It can be seen that there is no major difference between the models. Other distribution functions are also possible. The only reason for proposing a simpler model for the lowest accuracy categories is that it would be easier to implement.

10.2.2 Material model

Different material models used in CWM are described in Sections 9.2–9.10. The thermal properties, Young's modulus, E, and the thermal dilatation (expansion), ϵ^{th}, are the most important parameters, together with the yield limit. The latent heat due to melting is included in the thermal model. The plastic material model that is commonly used is the rate-independent, incompressible von Mises plasticity. It is combined with the associated flow rule and isotropic hardening. Dong *et al.* [190] recommends combined isotropic and kinematic hardening for

Table 10.4 Accuracy categories and heat input model

Basic and standard simulation	Constant flux in weld pool region
Accurate and very accurate simulation	Double ellipsoid heat source model

models that correspond to accurate or very accurate models in the proposal above. All these material properties are taken as temperature dependent.

These properties form the basis for a material model. There are three aspects of material modelling that need to be considered, in addition to the summary in Sections 9.2–9.10, when discussing the accuracy levels:

- The use of cut-off temperature, T_{cut}, discussed in Section 9.2. This is the temperature above which no further changes in material properties are accounted for in the mechanical analysis.
- Accounting for the effect of phase changes or other microstructural changes on the material properties as discussed in Section 9.2 and the following chapters.
- The consideration of rate-dependent plastic behaviour at higher temperatures (Section 9.7.1).

All models need to have good thermoelastoplastic properties up to the cut-off temperature. The use of cut-off temperature and the above listed issues for the plastic behaviour of the material modelling are related to the accuracy levels in Table 10.3. The homologous temperature is used in Table 10.5 where the material modelling issues for the different accuracy categories are summarised. It is defined as

$$T_{\mathrm{m}} = \frac{T}{T_{\mathrm{solidus}}} \qquad [10.1]$$

Table 10.5 Accuracy categories and material model

Basic simulation	$T_{cut} \geq 0.5T_m$ [a] Ignore volume changes due to phase changes but include their effect on the final obtained yield limit after cooling to room temperature. Temperature-dependent thermal properties and latent heat due to solidification are used.
Standard simulation	$T_{cut} \geq 0.7T_m$ [a] Include volume changes and other effects of phase changes on material properties. Temperature-dependent thermal properties and latent heat due to solidification are used.
Accurate simulation	Same as for standard accuracy.
Very accurate simulation	$T_{cut} \geq T_m$ It may be necessary to include rate dependency or even fluid flow into the model in order to have accurate high-temperature behaviour. May need a detailed model for the latent heat due to solidification as well as solidification shrinkage strain.

[a] Estimate based on A_{e1} and A_{e3} for low-alloy steels.

where T is the temperature and $T_{solidus}$ is the temperature at which the material starts to melt. The temperatures are given in Kelvin.

10.2.3 Geometry and discretisation

Discussions about the implications of using different subspaces of the general three-dimensional geometry and about choice of finite elements for spatial discretisation are given in Sections 4.3–4.4, 7.6, 7.8, 8.3 and 9.13. The choice of finite elements used in the CWM model constrains the simulated behaviour into a given space as given in Table 7.3. This table also explains the different notations for different kind of models used below, such as 2D-X. There is no simple relation between accuracy category in Table 10.3 and the choice of element formulation. It depends not only on the general scope of the simulation (Table 10.1), but also on what specific quantities are of interest. This can be illustrated by the work of Berglund and Runnemalm [178, 390], described in Section 7.6. For example, a 2D-X model cannot be used if the in-plane deformation of a butt-welded plate is needed. Choosing finite elements in the model requires an understanding of the fundamentals in continuum and structural mechanics and the expected deformation behaviour that is in the scope of the analysis. These fundamentals of solid mechanics are not covered in this book. Some remarks about choice of geometric models are given below.

The required computing power limits the use of three-dimensional solid elements for the higher accuracy levels as they are also related to the time and space discretisation. It is advisable to start with a two-dimensional model even if it is known that a three-dimensional model is needed. Large models with high accuracy require adaptive meshing (discussed in Section 7.3) and parallel computing (see Section 7.5).

The 2D-X models can, despite the approximations discussed in Section 7.6, be useful to obtain a residual stress state. An example of this is referenced below when discussing basic accuracy models. They can also be used for very accurate modelling when studying strain–stress behaviour near the weld for estimating risk for cracking and other local phenomena. This is because they can be combined with very small elements and time steps.

Two-dimensional plane stress, 2D-P, is useful when the in-plane deformations are of major concern. These kinds of model have been used to study gap changes in the weld joint and the effect of tack-welds on this or forces on tack-welds [71]. 3D-shell models are useful when simulating the welding of thin-walled structures in order to obtain the overall deformation behaviour and stresses. Then the details near weld are not of interest. It is possible to combine them with 3D solids near the weld in order to resolve the thermal and mechanical fields better in this region in order to study, for example, risk for cracking.

The work by Dike *et al.* [200] (see Section 7.8) is illustrative when considering the need for three dimensions. They needed a three-dimensional model to study the transverse deformation of a weld. They made attempts to improve their two-dimensional model to describe this deformation but it did not work: the changing restraint during the welding process affected the results too much.

It is possible to lump weld passes and ignore some of the thermal history for basic accuracy models as found by many investigators (see Section 7.7). Higher accuracy models need each weld to be modelled separately in order to obtain correct transient results. Accurate and very accurate models need many elements over the cross-section of the weld, whereas one element is sufficient for the standard accuracy model.

10.3 Motivation for proposed modelling strategy

10.3.1 Basic accuracy category

Most papers about finite element simulation of welding have focused on obtaining residual stresses [86]. Discussion of residual deformations is less common but there are still a considerable number of papers that include this aspect. Early papers by Ueda and coworkers [266, 267] investigated possible simplifications when simulating multipass welding. Two-dimensional models with simple material models were used with surprisingly good agreement with measurements. The use of temperature envelopes and/or lumping of welds is discussed in Section 7.7. Such techniques may work well but it is quite possible to perform complete analysis of large multipass welds using two-dimensional models [3] without them. The discussion about lumping, etc. is of more interest when using three-dimensional models.

The Satoh test can be used to discuss the influence of high-temperature behaviour on the residual state (see Lindgren [389]). It is assumed that the ends of the bar are fixed, i.e. an infinite large restraint. Then it is possible to estimate the temperature, T^y, from which the axial stress in the bar increases from zero to the yield limit of the material before cooling completely. An estimate gives, for steel with a yield limit of 600 MPa:

$$T^y = \frac{\sigma^y}{E\alpha} = \frac{600}{2 \times 10^5 \times 1.5 \times 10^{-5}} = 200\,°\mathrm{C} \qquad [10.2]$$

The influence of material properties above this temperature on the residual stresses is then not so large if a material has a low hardening coefficient. Thus it is possible to approximate, simplify or even ignore some parts of the material behaviour at higher temperatures as in the papers referenced in the beginning of this chapter. This temperature depends on the weld restraint [389]. The more rigid a structure is, the higher is T^y. This is the motivation for the geometry category *rigid* in Table 10.2. It can be simulated using a basic simulation accuracy level in Table 10.3 for obtaining good residual stresses. The note to

Table 10.3 is important if a material has a low martensite formation temperature [68] as this phase change may lead to a large reduction in the residual stress. A simple estimate whether the transformation temperature is 'low' or not can be done as follows. Assume that the martensite expansion completely removes the stress state. Then, if the martensite finish temperature is lower than T^y in Eq. [10.2] it affects the residual stress state and the geometry is classified as a *standard* geometry.

The smaller the weld is in comparison with the thickness of the joined material, the larger is the weld restraint. The papers by Ueda and Rybicki [191–193, 266, 267], where basic accuracy models were used, were all multipass weld applications.They used simple material models for multipass welds and obtained quite good agreement with measured residual stresses. Their material models had the characteristic of the basic accuracy level. Ueda and Yamakawa [161] introduced some effects of phase changes. But their model was a crude approximation with a cut-off temperature of 800 °C and the stress calculation was performed only for the last weld in each layer in the multipass weld procedure. Thus most of the history of the thermal cycling in the material was ignored. Therefore the analysis is also classified as a basic accuracy model. Mok and Pick [172] studied a T-joint made by 24 weld passes. They obtained quite good residual stresses compared with experiments and used a basic accuracy model. They only simulated the capping layer of welds. Kussmaul *et al.* [391] used a standard accuracy model for a multipass weld. They had quite an advanced model for accounting for the phase changes. Still they did not get good agreement with measured longitudinal residual stresses. This was not due to the material model but they were not consistent as their model simulated only the welding of the welds in the last layer. It may be that a basic accuracy model where all the welds were included would have given a better agreement with experiments.

Lindgren *et al.* [3] simulated a 28-pass weld where the effect of phase changes was ignored. This was a basic accuracy model. They obtained reasonable agreement with measurement. Börjesson and Lindgren [65] improved the material modelling by using an accurate TMM model. The residual stress state in the weld regions was changed even if there were no direct improvement when comparing with measurements. This change was due to the inclusion of phase changes but even more to the different material data used. The material properties in [65] were computed from mixture rules based on the microstructure evolution. Thereby this analysis had different material properties; in particular, the yield limit at the final room temperature. This explains the differences obtained compared with the basic accuracy model.

Thus it seems reasonable as a first estimate to classify multipass welds as rigid cases with a high weld restraint where only basic accuracy is needed if the residual state is the scope of the analysis. There are a large number of papers with this scope and accuracy level; see the review by Lindgren [86].

10.3.2 Standard accuracy category

Many of the publications in the basic accuracy category were published during the period 1970 to 1985 and focused on the residual state. Most of the models currently published belong to the standard accuracy category.

The one-pass weld studied by Tekriwal and Mazumder [258] should be analysed with a standard accuracy model to obtain the residual stress state according to the proposed modelling strategy. They (see Section 9.2) varied the cut-off temperature from 600 °C and up to the melting temperature. However, contrary to the current guidelines, they did not find any large effect of cut-off temperature on residual stresses. This is because they did not include phase transformations in their model. Thus there were no larger changes in the material data when lowering the cut-off temperature from 900 to 600 °C. There would have been larger changes in the residual stress field if they had done so. This was the case in the study by Oddy *et al.* [316]. They investigated the effect of including more or less of the influence of the phase changes on the material behaviour. They investigated a one-pass weld and found a large influence on the residual stresses (Fig. 9.16). This is in accordance with the current proposal. They showed that the ignorance of these effects was the reason for the discrepancy between measurements and simulations for the one-pass weld in the paper by Hibbitt and Marcal [127]. Lindgren [392] repeated the simulation by Andersson [88]. It was a plane deformation model of a one-pass weld butt joint. Andersson used a standard accuracy model accounting for phase transformation effects on the material properties. He obtained good agreement with measurements. Lindgren varied different aspects of the material model and obtained only minor changes on the residual stresses near the weld (see Fig. 10.2 together with Table 10.6 for an explanation of the different variations).

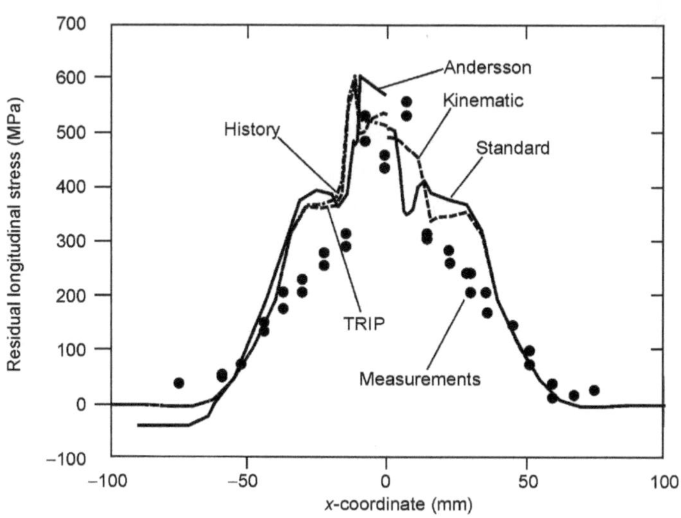

10.2 Residual stresses on upper side of plate [88, 392].

Table 10.6 Different material models used for results in Fig. 10.2

Notation	Description
Standard	Two different materials, one for weld and one for filler. Only temperature-dependent data.
Andersson	Same as *Andersson* [88] which gives different thermal expansion during cooling.
History	As *Andersson* but also different yield limit during cooling.
Kinematic	Same as *history* but with kinematic hardening.
TRIP	Same as *history* but with transformation plasticity.

Lindgren and Karlsson [164] studied the butt-welding of a thin-walled pipe (Fig. 4.20). It was a one-pass weld. They used an almost standard accuracy model (only transformation plasticity was missing), and comparisons were made by Lindgren in [86] with other studies [75–77]. It was found that changes in the material model affected the residual stresses in the weld region. Thus a standard accuracy model is required. The good agreement with measurements at the weld centre in the paper by Murthy *et al.* [77] seems to be due to better data for the thermal dilatation during cooling.

McDill *et al.* [165] (see Section 9.2) simulated a one-pass weld on the edge of a slender beam. This case can be considered as a standard or flexible geometry (see Table 10.2). A standard accuracy model was used, which gave quite good agreement for the SS steel but not for the MS steel. The latter discrepancy was believed to be due to inaccurate material properties, as they did not match the actual properties of the carbon steel bar.

10.3.3 Accurate and very accurate categories

More advanced material models, such as TMM models with finer meshes, correspond approximately to accurate models; see earlier chapters for details. This kind of model is becoming common in two-dimensional studies. Several of the publications use these models. There is no distinct line between standard and accurate models; it depends on the accuracy requirement as well as how much detail near the weld needs to be resolved.

The major extension for the very accurate category is the use of viscoplasticity in the material modelling. This is a distinct difference in the modelling needed. Simultaneously, there is a need to have models for hot or solidification cracking when these issues are the scope of the simulation. Some of the references in Chapter 12 address this. The requirement on the model in case of flexible structures is also under development (see Sections 4.3 and 8.5.4).

11
Robustness and stability

There are variations in process parameters, geometry and material properties even when the same nominal values are attributed to them. These variations are statistical and define a process window. It is important that variations within this process window do not lead to changes in the geometry outside tolerances or a shorter than expected lifetime of the welded component. There are methods, discussed in the current chapter, to use simulations to investigate whether a process is robust or not. The underlying assumption is that a numerical validated stable model exists so that possible numerical problems are due to physical phenomena and not to a badly created model or chosen simulation parameters such as time stepping and tolerance criteria. Numerical instabilities are caused by the numerical algorithm and the discretisation of the model equations, whereas the physical instabilities are due to the stability of the solution of a model. Very little is known about how stable numerical procedure behaves in physically unstable processes (Section 6.6 in [34]).

The underlying concepts with respect to modelling were defined in Chapter 9 and those relevant for validation of models were discussed in Chapter 8. Instability is considered as an extreme case of a non-robust process. Different methods of parameter variations are appropriate for instability whereas there are specific numerical techniques to determine whether different states are stable or not. The question of robustness and stability is mainly concerned with the mechanical behaviour in CWM although some of the discussions and notations below are general.

11.1 Definitions concerned with robustness and stability

There is no clear borderline between the concept of robustness and stability. It is possible to consider instability as the ultimate example of a process that is not robust. However, it all comes down to the relation between small changes in input of a process on its output. Belytschko and Mish [217] call these variations in the input for irreducible errors in the computational model, as they are due to stochastic variations that are not controllable. They are also called aleatory uncertainties [213]. The notions of robustness and stability are used in dynamics and control theory.

Stability is also connected with the case of multiple solutions where some may be stable and other unstable. The theory of stability originates from Lyapunov [34, 393]; Lyapunov stability occurs when small perturbations of a stable solution remain small. A more precise definition is based on the definition of a metric. This metric is used as a measure of when the difference between two solutions is small. This can be quite natural in engineering problems. The metric is then typically the size of some deformations. More general metrics can be defined and they should all have the same property as the metric 'distance' has in geometry (Chapter 7 in [393]). Adapting this concept to finite element solutions gives the metric (Section 6.5 in [34])

$$s(t)_{\mathrm{AB}} = \|\boldsymbol{d}(t)_{\mathrm{A}} - \boldsymbol{d}(t)_{\mathrm{B}}\| \qquad [11.1]$$

where s_{AB} is the 'distance' between solutions A and B at time t, $\boldsymbol{d}$ is a vector of finite element solution variables and the norm is typically the L_2-norm. It is possible to have different types of variables derived from the finite element solution in this measure. The definition of a process that is Lyapunov stable if a perturbation of the initial state

$$s(0)_{\mathrm{AB}} = \|\boldsymbol{d}(0)_{\mathrm{A}} - \boldsymbol{d}(0)_{\mathrm{B}}\| \leq \epsilon \qquad [11.2]$$

results in

$$s_{\mathrm{AB}} = \|\boldsymbol{d}_{\mathrm{A}} - \boldsymbol{d}_{\mathrm{B}}\| \leq C\epsilon \qquad \text{for all } t > 0 \qquad [11.3]$$

A process is asymptotic stable if

$$s(t)_{\mathrm{AB}} = \|\boldsymbol{d}(t)_{\mathrm{A}} - \boldsymbol{d}(t)_{\mathrm{B}}\| \to 0 \quad \text{for all } t \to \infty \qquad [11.4]$$

Different methods exist to evaluate this, as discussed below. The concept of Lyapunov stability is also used to evaluate the stability of numerical algorithms [23, 393] either by checking that rounding off errors, the perturbations, do not grow or by using the concept of a Lyapunov functional, $\Pi(\boldsymbol{d})$, as in Armero and Simo [23]. This functional should be positive definite with respect to the metric and have an upper bound for stable process. Furthermore, it should be non-increasing function. This variant will not be discussed below. The importance of the metric and the Lyapunov functional above should be noted. A process can be stable with respect to one metric but not the other.

Robustness is defined the same way in control theory. There it is stated that a process is robust with respect to parameters in $\boldsymbol{p}$ and a given metric. Thus the parameters in $\boldsymbol{p}$ correspond to the disturbance of the initial state in Eq. [11.2]. A first step when investigating the stability of processes is this parameterisation of the problem. With respect to what parameters should the stability of the system be checked? Typically, this is the load of a system. The magnitude of the disturbance, the load, is varied by introducing a control parameter λ so that the 'loading' becomes $\lambda \boldsymbol{p}$. A set of solutions for varying λ exists and can usually be grouped into branches. There are different kinds of critical points along these branches where the bifurcation point is of particular interest in the current context. Multiple solutions occur at these points that lead into different branches. Those are stable or unstable.

The stability of the equilibrium solution of the simple deformation case is illustrated in Box 11.1. The stability analysis is demonstrated by the potential energy of the system, and by linearised buckling analysis.

Box 11.1 Vertically loaded rigid rod with torsion spring

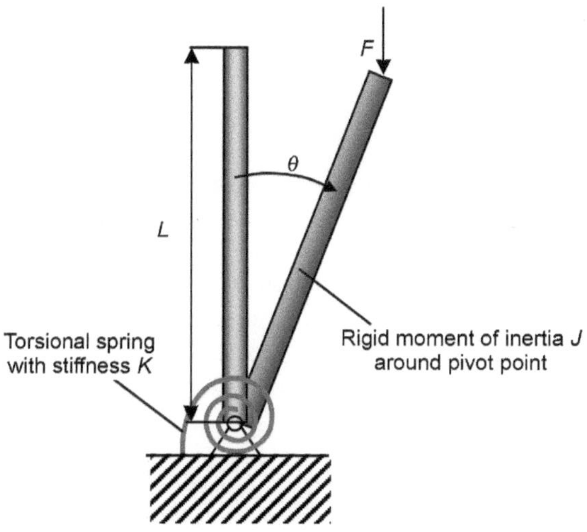

11.1 Vertically loaded rigid rod.

A vertically loaded rigid rod is shown in Fig. 11.1. The solution of the equilibrium equations accounting for large rotations is included as reference. Moment equilibrium around the pivot point and ignoring gravity, accounting for large rotations, gives

$$K\theta_{\mathrm{eq}} - FL\sin(\theta_{\mathrm{eq}}) = 0 \qquad [11.5]$$

One solution is the trivial zero solution. Another non-zero solution is also possible when

$$F > \frac{K}{L} = F_{\mathrm{cr}} \qquad [11.6]$$

It can be shown that the trivial solution is stable until the load reaches the critical value in Eq. [11.6]. Thereafter it becomes unstable and the deformed configuration is the stable one. See Fig. 11.2.

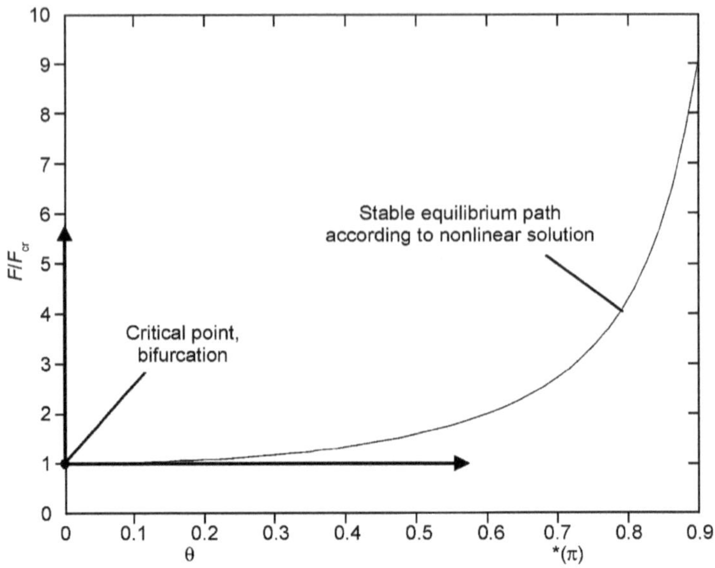

11.2 Equilibrium paths and critical point. $\theta = 0$ is also one solution and it is unstable when $F/F_{cr} > 1$.

A linearised buckling analysis is given below. The stability of the vertical position of the rod for different loads is investigated. The theory for buckling analysis is summarised in Section 11.3. A control parameter λ of the load is introduced as a perturbation of the load. The static criterion is applied on the undeformed geometry in order to find critical points. Equilibrium gives

$$K\Delta\theta - \lambda FL\Delta\theta = 0 \qquad [11.7]$$

or

$$(K - \lambda FL)\Delta\theta = 0 \qquad [11.8]$$

It is possible to use the finite element notations in Chapter 6 calling K for the stiffness matrix. The derivative of the displacement dependent force with respect to the angle takes the role as geometric stiffness matrix. Then we can write

$$(\boldsymbol{K}_c - \lambda\boldsymbol{K}_\sigma)\Delta\boldsymbol{\theta} = \boldsymbol{0} \qquad [11.9]$$

The non-trivial solution is possible if

$$(K - \lambda FL) = 0 \quad \text{or in matrix form} \quad \det(\boldsymbol{K}_c - \lambda\boldsymbol{K}_\sigma) = 0 \qquad [11.10]$$

The solution of λ for this characteristic equation is

$$\lambda_{cr}F = F_{cr} = \frac{K}{L} \qquad [11.11]$$

There is only the trivial solution $\Delta\theta = 0$ except at λ_{cr}. This is the critical point where multiple solutions exist. The trivial and multiple solutions are illustrated by the arrows in Fig. 11.2.

A Lyapunov functional can be used to check for critical points also. This is the potential energy of the system in this case

$$\Pi = \frac{1}{2}K\theta^2 - \lambda FL[1 - \cos(\theta)] = 0 \qquad [11.12]$$

It should be positive definite at the stable equilibrium points. The equilibrium angle can be found from Eq. [11.5] or by the condition of stationary potential energy

$$\frac{d\Pi}{d\theta} = K\theta_{eq} - \lambda FL\sin(\theta_{eq}) = \lambda FL\left[\frac{F_{cr}}{\lambda F}\theta_{eq} - \sin(\theta_{eq})\right] = 0 \qquad [11.13]$$

This gives the same equation as in Eq. [11.5]. The potential energy is positive definite, i.e. the solution is stable, if the second derivative is positive. We have

$$\frac{d^2\Pi}{d\theta^2} = K - \lambda FL\cos(\theta_{eq}) = \lambda FL\left[\frac{F_{cr}}{\lambda F} - \cos(\theta_{eq})\right] \qquad [11.14]$$

The solution $\theta_{eq} = 0$ is stable when $\lambda F < F_{cr}$ as then the value is positive, otherwise it is negative when $\lambda F > F_{cr}$. There is also a nonzero equilibrium solution for the latter case. Equations [11.12] and [11.13] can be combined to give

$$\left.\frac{d^2\Pi}{d\theta^2}\right|_{\theta=\theta_{eq}} = \lambda FL\left[\frac{\sin(\theta_{eq})}{\theta_{eq}} - \cos(\theta_{eq})\right] \qquad [11.15]$$

which is positive for $\theta_{eq} \in [0, \pi]$. Thus this is stable at loads above the critical value.

The dynamic criterion to find critical points is similar to the static criterion. The dynamics, inertia, of the system is included. Then the equation of motion becomes

$$J\ddot{\theta} + K\theta - \lambda FL\theta = 0 \qquad [11.16]$$

A solution of the type $\theta = Ae^{rt}$ is assumed. Inserting this into above gives

$$(K + r^2J - \lambda FL)Ae^{rt} = 0 \qquad [11.17]$$

non-trivial solution when the characteristic equation is zero (i.e. the determinant is zero), critical point

$$r = \pm\sqrt{\frac{\lambda FL - K}{J}} \qquad [11.18]$$

Imaginary roots correspond to harmonic vibrations whereas real roots will give growth terms in the solution and thus an unstable solution. This is

$$\lambda F > \frac{K}{L} = F_{cr} \qquad [11.19]$$

as before.

A remark can be made that the trivial, zero angle, solution when the force is above the critical value is not unstable in Lyapunov sense if the rotation is used as a metric as it is always less than π.

11.2 Perturbation methods for investigation of robustness

The statistical design of experiments (DOE) [394, 395] is a method for planning and conducting experiments when investigating relations between input and output to a process. When optimising a process the aim is to determine the region in the important factors that leads to the best possible response. Another possible objective of an experiment is to test the robustness of a process, that is, under what conditions do the response variables of interest degrade or how can the variability be reduced in the response variable that arises from external sources?

First design factors, i.e. influencing parameters, have to be selected. Thereafter the number of levels, values, for each factor and the specific levels at which runs will be made must be chosen. If the objective of the experiment is factor screening or process characterisation, the number of factor levels should be held low. Generally, two levels work very well in factor screening studies. The ranges or region over which each factor will be varied must also be chosen. In factor screening the region should be large. When the experimenter learns more about which factors are important and which levels give the best results, these regions will become narrower. When selecting the response variable, or the output response, the experimenter has to be certain that this variable really provides useful information about the process under study. Multiple responses can occur. A response variable may be discrete or continuous. To organise the generated information, the cause-and-effect diagram or the fishbone diagram can be useful. In this diagram the response variable of interest is drawn along the spine of the diagram and the design factors are organised in a series of ribs.

The next step is the choice of experimental design. There are different types of experimental designs, one of the most important and widely used in industrial experimentation are the factorial designs. These designs investigate the effects of two or more k factors, each at two levels; the design is called a 2^k factorial design. In these designs the factors are varied together instead of one at a time, which means that possible interaction between the factors can be observed. Such interactions cannot be observed in a one-factor-at-a-time approach. If the number of factors is increased, then the total number of experiments will become very high since it increases exponentially. To reduce the number of runs it is possible to use a fractional factorial design where only a part of all the possible runs is performed. This method is widely used when performing screening experiments.

DOE has been used in welding contexts. Olabi *et al.* [396] aimed at finding the relation between some weld configuration parameters and the residual stress in the HAZ. Thus the objective of the 'experiment' is the stress in the HAZ. First influencing parameters (factors) are chosen and their ranges (levels). Thereafter, a polynomial relation between the stress and these parameters is set up. Following the DOE methodology, a set of 'experiments' is determined. This is, in this case, a number of simulations with varying parameter values. The final analysis of the results yields a simple formula relating the stress with the chosen factors. Casalino *et al.* [397] use a trained artificial neural network (ANN) as virtual laboratory in a DOE approach.

11.3 Methods for analysis of stability

There are two different kinds of approach for checking the stability of a process with respect to the parameters that are chosen as perturbations of the process [93]. The direct methods aim at finding the critical points directly by solving some suitable sets of equations. This may be a nonlinear set of equations requiring good initial guesses in order to work. The alternative, indirect methods, is to detect them when traversing an equilibrium path. The methods below are illustrated in Box 11.1.

The direct methods are based on extending the original residual forces equations with a set of constraints that define a critical point [398]. The standard incremental, iterative solution procedure, Eq. [6.23], is written as

$$ {}^{n+1}_{i}\boldsymbol{K}_t(\lambda)_i \delta \boldsymbol{U} = {}^{n+1}_{i}\boldsymbol{R}(\lambda) \qquad [11.20] $$

where we have introduced the control parameter λ in order to emphasise that this is to be solved at a certain load level. Equilibrium exists when ${}^{n+1}_{i}\boldsymbol{R}(\lambda) = \boldsymbol{0}$. Note that there are terms in the tangent matrix that also depend on the load level. Wriggers *et al.* [398] proposed that the instability is characterised by the second derivative of the potential energy becomes negative and expressed this by the condition

$$ \det\lfloor {}^{n+1}_{i}\boldsymbol{K}_t(\lambda) \rfloor = 0 \qquad [11.21] $$

Positive determinant holds for a stable path whereas negative for an unstable path. The extended problem is then

$$ \begin{bmatrix} {}^{n+1}_{i}\boldsymbol{R}(\lambda) \\ \det\left[{}^{n+1}_{i}\boldsymbol{K}_t(\lambda)\right] \end{bmatrix} = \boldsymbol{0} \qquad [11.22] $$

Wriggers *et al.* suggested different ways of solving this extended problem. The linearised buckling analysis is one example where the problem above is decoupled. The evaluation of the stiffness matrix takes place at a given deformation or for the undeformed geometry. The tangent stiffness matrix has two contributions: the constitutive, or material, matrix and the geometric stiffness matrix; Eq. [6.33]. The latter is sometimes called the stress stiffness matrix. The latter part is assumed to depend linearly on the control parameter so we can write

$$\boldsymbol{K}_t = \boldsymbol{K}_{\mathrm{c}} + \lambda \boldsymbol{K}_\sigma \qquad [11.23]$$

The instability occurs when a non-trivial (non-zero) solution of

$$(\boldsymbol{K}_{\mathrm{c}} + \lambda \boldsymbol{K}_\sigma)\boldsymbol{v} = \boldsymbol{0} \qquad [11.24]$$

exists. We can rewrite this as

$$\boldsymbol{K}_{\mathrm{c}}\boldsymbol{v} = -\lambda \boldsymbol{K}_\sigma \boldsymbol{v} \qquad [11.25]$$

which is a generalised eigenvalue problem. The eigenvalues found, λ_n, means that a non-trivial solution exists, i.e. instability occurs, for the load levels λ_n multiplying the current load or stress level. It should be noted that it is the current stress state that is perturbed, not for example welding speed and other process variables. The lowest value of $\lambda_{\min}$ is the important one. If the control parameter is near unity, then the current load is near buckling. The linearised buckling analysis is based on assuming conservative loading and elastic material behaviour. The matrices are evaluated at the undeformed geometry in the direct approach. Thus it corresponds to assuming that prebuckling deformations are small.

It is possible to evaluate the matrices at a given deformed state. Then the analysis requires a simulation that follows an equilibrium path along which the analysis above is repeated at different occasions. Thus this approach belongs to the category of indirect methods for determination of critical points. The linearisation around the deformed state is based on the same assumptions as given above for the direct variant of the method except that the prebuckling deformations are not assumed to be small. The indirect variant is relevant for CWM models and can give a good estimate of the buckling loads [34]. One limitation is that the current tangent stiffness matrix does not describe the behaviour of the system for those regions that may elastically unload for the relevant buckling mode.

The eigenvalue analysis approach described above is a static criterion for buckling. The dynamics of the structure is not included in the analysis. This may be necessary to account for as the buckling can be very fast and inertia may affect the behaviour in the same manner as the rotational inertia included in the rod in Box 11.1. A more general dynamic criterion can be formulated: then the mass matrix must also be included. The eigenvalue analysis gives characteristic exponents and associated modes of vibrations, eigenmodes. The characteristic exponents are complex numbers and if none of them has a positive, real component, then the structure is dynamically stable. The exponents depend on the load control parameter as it is varied.

11.4 Application of robustness and stability analysis in CWM

A parametric study can be used to ensure the robustness of a manufacturing process. Usually the focus is on keeping the dimensions within the specified tolerance. Mechanised welding reduces the variations in the welding process itself. However, the variations in the incoming material can cause unexpected changes in the deformation behaviour.

A large number of papers include variations in the material properties and their effect on the computed results. The aim has usually been to investigate not the robustness of the welding process but rather the robustness of the model itself with respect to uncertainty in material parameters [165, 215, 297, 399–402]. Michaleris [402] performed a material parameter variation to study the effect on the results.

Eigenvalue analyses are unsual in CWM. Buckling behaviour of a weld configuration is often revealed in a normal welding simulation when large deformations are accounted for. The risk for buckling is then seen either by very large computed deformations or by convergence problems. The latter may be due to numerical issues but may also have real physical causes. Poorly conditioned matrices and/or convergence problems occur when small changes in the load give large changes in displacements. Therefore, convergence problems may indicate the presence of an instability. An eigenvalue analysis can be made just before the instance in a simulation when convergence failure occurs. This eigenvalues analysis may show whether it is a buckling tendency. The eigenmode can indicate how the welding procedure should be redesigned.

Michaleris and coworkers [402–404] performed a buckling analysis on the welded structure. This was done by a large deformation analysis and not by an eigenvalue analysis. Troive [78] investigated the axial buckling strength of a pipe with a curcumferential weld. He used a large deformation analysis and calculated the axial load–deflection curve to find the buckling strength. Thus he did not perform an eigenvalues analysis. He found that welding residual deformations have a larger effect than residual stresses on the buckling limit of a structure.

12
The current state of computational welding mechanics (CWM)

The three-part review by Lindgren [86, 209, 292] covers most published work up to 2000 and some additional references are given in Lindgren [379]. Mackerle [405, 406] lists references to work in CWM. Development since 2000 has been both in terms of improved as well as larger models. Some of these papers have been referred to earlier in this book. Below a selection of recent papers is discussed. They illustrate the current developments of model, software as well as industrial applications. The work by Goldak *et al.* [407] illustrates the future of enabling designers, i.e. non-finite element analysis specialists, to perform modelling and simulation in CWM (see Fig. 12.1).

12.1 Heat input models

The work by Lu *et al.* [408] does not include the mechanics of welding but is still of interest in this context. They formulated an integrated model of the arc plasma and weld pool (Figs 2.1 and 2.2), in order to find the energy distribution and weld pool geometry. Hongyuan *et al.* [409] extended the double ellipsoid heat input model with arc perturbation angles.

12.2 Material models

Thiessen and coworkers [333, 410, 411] combined the welding simulation with a phase field method for the microstructure calculation. The phase field domains were residing at the nodes and occupied only a fraction of the elements in order to reduce the computational effort. However, these domains were close enough to give a semi-continuous representaion of the microstructure. They used a thermal model of the welding process with the double ellipsoid heat source for the thermal analysis. The temperature field drove the phase field calculations and no simulation of the mechanical behaviour was done. The microstructure was compared with synchrotron measurements.

Elmer *et al.* [412] and Zhang *et al.* [413] studied the welding of AISI 1005. They applied the JMAK model (Section 9.9.2) for austenite formation and a

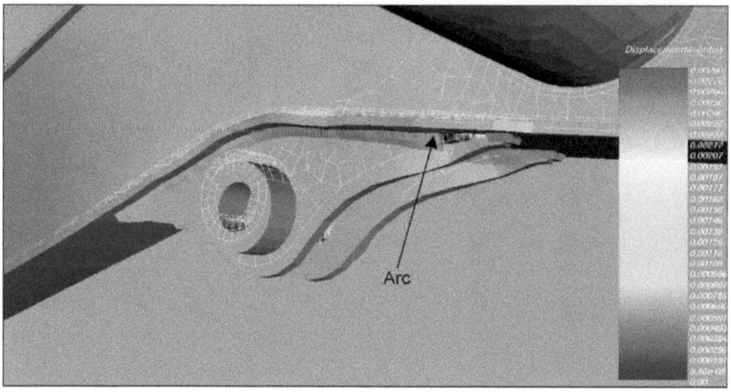

12.1 Welding of feature on boom [407].

model by Bhadeshia *et al.* [414] for its decomposition to α-ferrite again. This model was developed for low-carbon steels. They computed austenite grain growth using Monte-Carlo simulations. *In-situ* X-ray diffraction techniques using synchotron radiation (SRXRD) for determination of phase maps and measurements of the HAZ microstructure after welding were performed [412]. The *in-situ* measurements [415] opened up new possibilities to validate microstructure models for welding applications. Jensen *et al.* [416] also used synchrotron measurements for model validation.

Bergheau *et al.* [417] used an axisymmetric model of welding with a viscoplastic material model. This is a very accurate model according to the classification in Table 10.3 and description in Table 10.5. They used the TRIP model by Leblond (Section 9.7.2) adapted to viscoplasticity [418]. Their purpose was to evaluate the effects of high-temperature viscous behaviour on the residual stresses and deformations. The experimental set-up welded a small, circular disc. The 'small' radius was chosen in order to maximise the influence of the high-temperature phenomena on the response. In this sense, the structure may be called a flexible structure according to Table 10.2. They used a TMM model with the microstructure model according to Leblond (described in Section 9.9.1), and the classical Koistinen–Marbuger equation for martensite formation (see Section 9.9.4). They found considerably better agreement when using the viscoplastic model than when using a rate-independent plasticity model. They used data for the latter based on low strain-rates that resemble those during cooling but lower than those during heating.

Vincent *et al.* [69] performed Satoh tests of varying kind with multiple thermal cycles. These test were used to evaluate constitutive models and material behaviour. The tests form a basis for investigation of the TRIP effect (Section 9.7.2), viscoplastic material behaviour (Section 9.7.1) and memory effects (Section 9.3). Preston *et al.* [419] simulated welding of aluminium and accounted for the microstructure evolution along the same lines as Grøng [60] and validated the model versus X-ray synchrotron diffraction measurements.

12.3 Models for fatigue and cracking

Barsoum [420] simulated the welding of a plate to a pipe and validated the model with measurements. The residual stress field was used as initial stresses in a simulation where a sinusoidal varying torsional moment around the centre axis was applied. The work was part of a project for fatigue assessment of welded structures.

Wei *et al.* [421] studied liquation cracking assuming that the driving force is the strain versus temperature at the trail of the weld pool. The results were compared with measurements. They also simulated the trans-varestraint test and also monitored the strain in the brittle temperature range of the material.

Dye *et al.* [422] used the steady-state analytic solution for the temperature field due to a point source and a plane stress model in the mechanical analysis. The latter was applied in an Eulerian frame (Fig. 3.13), and thus a steady state analysis was done. They combined this simple model with detailed, advanced models for the metallurgical phenomena in order to predict formation of weld centreline, solidification cracking and liquation cracking.

Teng *et al.* [423] used a 2D-X model for an integrated analysis according to the definition by Radaj [424]. Computed residual stresses due to multi-pass welding were combined with in-service loads to calculate stress and strain in critical areas. This gives a strain-life curve used to compute the fatigue damage that is accumulated until a critical value is reached. This predicted lifetime is compared with measurements and the proposed methodology was found better than standard specifications for fatigue life. The approach was also used in Teng *et al.* [425].

12.4 Computational efficiency

Duranton *et al.* [426] used a 3D-solid model to simulate the welding of a 13-pass weld. It was a circumferential weld in a pipe made of a stainless steel. They used a hierarchical remeshing technique as in the right part of Fig. 7.1. Their remeshing is a dynamic remeshing as it seems that they prescribed the refinement and coarsening of the mesh and did not use an error measure to guide the remeshing.

12.5 Simplified methods

Audronis and Bendikas [427] compared several analytical solutions for deformations, 2D-X and 3D-solid finite element solutions with experiments. The case was a small plate welded on top of another plate. They were made of a stainless steel. However, the analytic solutions varied considerably, and the finite element solutions did not agree well with measurements. They indicate that this could be due to lack of material properties. Furthermore, the weld was laid instantaneously in the 3D-solid model.

Camilleri and coworkers [428, 429] developed simplified approaches for finite element simulation of welding. Their work included measurements. The first paper [428] compared a number of variants of an inherent shrinkage model (Section 7.8). The second paper used the most efficient variant of the models from their first paper and compared them with a 'complete' thermo-elastoplastic simulation. They found that the simplified model gave a good indication of out-of-plane deformation and proposed to use this for initial assessment whereas the 'complete' model could be used for a more detailed prediction in later design stages.

12.6 Reducing risk for buckling, deformations or residual stresses

Deo and coworkers [403, 404, 430] used simulations to study different approaches to reduce the risk for buckling of a welded structure. Van der Aa *et al.* evaluated low-stress-no-distortion (LSND) techniques, a trailing heat sink after-arc, applied on a butt-welded plate [431]. They developed a conceptual model for the mechanism.

12.7 Improved life

Toyoda and Mochizuki [432] investigated the effect of heat input and interpass temperature on ultimate strength and fracture toughness for a welded steel joint. A large number of tests were performed to evaluate this. They used a 3D-solid model and computed microstructure as well as Vickers hardness. They assigned thermal properties to each phase and applied a mixture rule for them. The focus in their evaluation is on the Vickers hardness and this depended only on the material composition, microstructure and the cooling rate. Thus the mechanical analysis did not play any role in their evaluation.

12.8 Repair welding

McDonald *et al.* [433] studied repair welding. A comparison between measured residual stresses on the surface and computed values was used to assess whether it would be sufficient in future similar repair welds to draw conclusions only on the internal residual stresses from surface measurements.

Elcoate *et al.* [434] performed simulations using three-dimensional models (with some symmetry assumptions) for varying repair procedures.They ignored the residual stress field from the initial weld and assumed that the stresses due to the local repair weld dominate. Furthermore, they ignored the transient behaviour during the repair welding, i.e. each repair weld was laid instantaneously. Comparisons were made with measurements and a fair agreement was obtained.

Vinas *et al.* [435] simulated a letter-box repair weld in the bottom dome of a boiler in a power station. The repair region could not be post-weld heat treated and therefore simulations and measurements were used to determine the effect of the repair weld. They used both two- and three-dimensional models. The outcome was a repair procedure that had lower stresses on the inner surface of the boiler.

12.9 Optimisation

Optimisation methods need the gradient of the solution with respect to the design variables. The common approach is to use parameter perturbation to find the gradient of the solution and search for the optimum. Analytical gradients are more efficient than the numerical approach. Michaleris and coworkers have developed sensitivity equations in CWM. The paper by Song *et al.* [436] is a further step in this development. The sensitivity equations are implemented in a three-dimensional formulation based on the Lagrangian reference frame and using a radial-return stress update algorithm. Earlier work was implemented in an Eulerian frame [44].

Optimisation of weld sequences is a combinatorial optimisation problem. An approach for reducing the computational effort for this case was applied on an aero engine component by Voutchkov *et al.* [437]. They utilised surrogate models to reduce the necessary computational effort. The surrogate model was 'trained' by the full computational model to represent the desired relationships that are of interest in the optimisation.

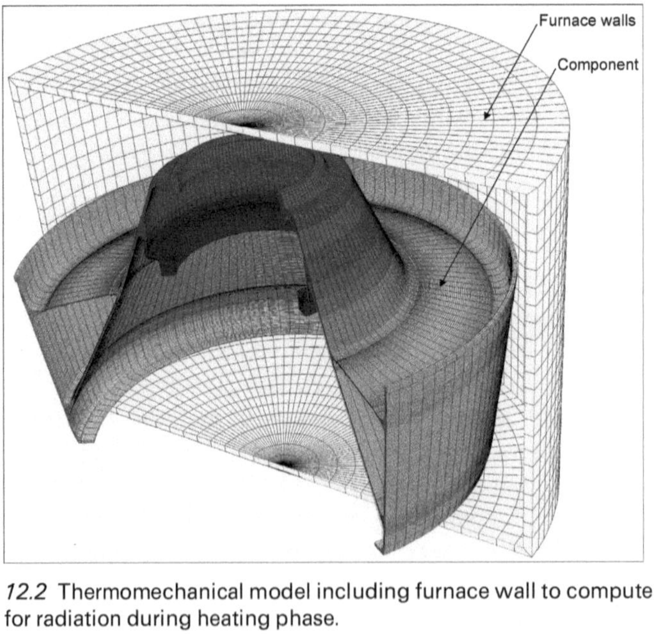

12.2 Thermomechanical model including furnace wall to compute viewfactors for radiation during heating phase.

12.10 Simulation of welding and other manufacturing steps

Berglund *et al.* [438] and Berglund and Alberg [72] simulated welding and subsequent heat treatment of an aero engine component. They also performed a gas flow analysis of the furnace to obtain boundary conditions during the cooling phase in the heat treatment simulation (see Figs 12.2 and 12.3).

Cho *et al.* [439] also simulated welding and subsequent heat treatment. Thorborg *et al.* [440] simulated welding, removal of material and heat treatment in order to investigate premature rupture that occurred in the weld zone.

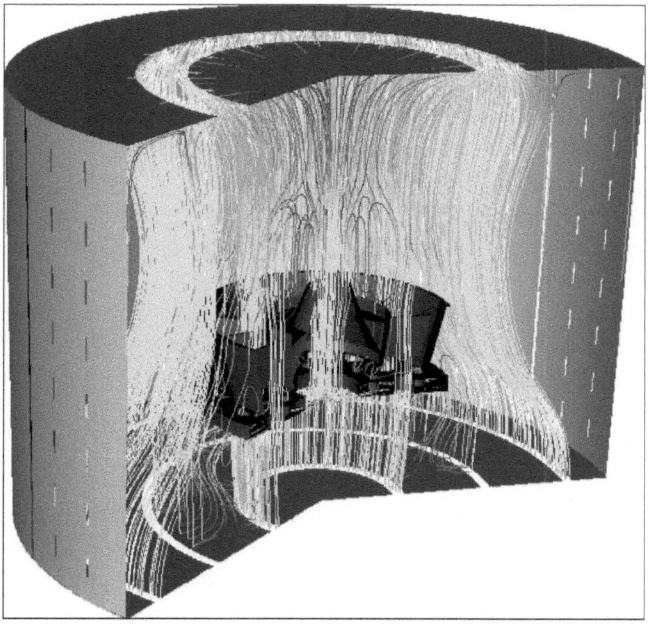

12.3 Stream lines during flow of cooling gas of welded aero-engine component.

13

The Volvo Aero story in computational welding mechanics (CWM)

H RUNNEMALM and H ALBERG, Volvo Aero Corporation, Sweden

13.1 History of CWM at Volvo Aero Corporation

Computational welding mechanics (CWM) emerged as an important technique for solving industrial welding problems during the late 1980s. CWM simulations were then mainly confined to the academic environment. However, the management at Volvo Aero had already recognised that CWM was very relevant to their needs. As a result, several research and development projects were undertaken together with Luleå University of Technology during the following decade. The expertise accumulated from this experience, together with developments in modelling techniques and computer power, made it possible to begin applying CWM to solve large-scale industrial problems. The first weld simulation carried out in-house at the company was undertaken in 1999. This was at the time when three-dimensional models had just started to appear in the research literature. One milestone for Volvo Aero was reached during 2002 when the first complete three-dimensional model of a component with more than 10 metres of welds was completed. This stage also included physical trials to validate the model. The period between 2003 and today has been dedicated to further development of CWM tools to enhance modelling performance, developing a database of weld performance, and establishing appropriate validation and optimisation strategies. However, as in all industrial environments, the main effort has been dedicated to using CWM in product development and manufacturing support. CWM is now used daily to ensure high-quality production of fabricated structures as well as repair welding.

13.2 Strategic decisions for successful implementation

The core of the Volvo Aero strategy to develop CWM was born in the vision 'What we cannot simulate we will not do' formulated nearly two decades ago. Several practical issues needed to be addressed before CWM simulations could be introduced into the organisation:

- What software can be used? Is commercial software available or must in-house software be developed? Is it necessary to use specialised CWM software or can a general-purpose finite element code be used?
- What competences are needed?
- Where in the organisation should CWM competences be located? Within the manufacturing division or the design department?
- Who needs to be involved in the development process?
- Where in the global development process (GDP) for Volvo Aero should CWM be introduced?

The following decisions were taken at Volvo Aero in response to the above issues:

- Selection of general-purpose software as a platform for continuous improvement. Particularly important was an open architecture that allowed user enhancement through coding of new algorithms.
- Close collaboration with the research community in development of specialised competences in CWM together with in-house expertise ranging from materials engineering, solid mechanics to production engineering.
- The location of CWM expertise in the manufacturing divisions to ensure a close connection with those who 'own' the welding processes such as welding engineers and operators.
- The provision of management support throughout the organisation. Implementation involved a wide range of staff such as welding engineers, weld operators, designers, project managers, those responsible for testing, material specialists. The process even involved the marketing department as well as the in-house staff newspaper published by Volvo Aero.

The timing of the introduction of CWM modelling was fortunate. The availability of the right expertise and technology ensured that rapid progress was obtained, making the process and product development process more efficient and competitive.

13.3 Business motivation for CWM

Aero-engine components are made of expensive materials. The geometrical tolerances on them are narrow and at the same time there is a demand for reduced product and manufacturing development time, i.e. reduced costs. This has led to an interest in fabricating these components instead of producing them from large castings. Many aerospace components with complex geometry can then be fabricated from smaller parts using joining techniques such as welding. However, the welding usually results in unwanted deformation and stresses. It is therefore important, given safety and cost requirements, to be able to reduce these unwanted effects on the components before attempting to fabricate them.

The risks and costs associated with development of these components has led to an increasing need to do things 'right first time'. This has in turn motivated development of modelling and simulation techniques that can reliably predict the effect of manufacturing processes such as welding on a product. Using computational models makes it possible to decrease risk and costs by increasing the information available about a product and its manufacturing processes. This will in turn strengthen the competitiveness of the company in the market.

Manufacturing simulations can be used when developing both products and associated manufacturing processes, with either new or existing products and also when considering investment in new equipment. Manufacturing simulations link design and manufacturing during product development and act as a tool for design and manufacturing engineers to evaluate different concepts or manufacturing processes.

At Volvo Aero, the design part of the product development process is divided into three stages:

- concept design;
- preliminary design;
- detailed design.

The manufacturing part of the product development process is also divided into three stages (Fig. 13.1):

- inventory of known methods;
- preliminary preparation;
- detailed preparation.

CWM can be used in all stages of product and manufacturing development. However, it is crucial that the methods and models used to carry out the manufacturing simulation are developed and implemented in the product development process before the simulations need to be used in live development

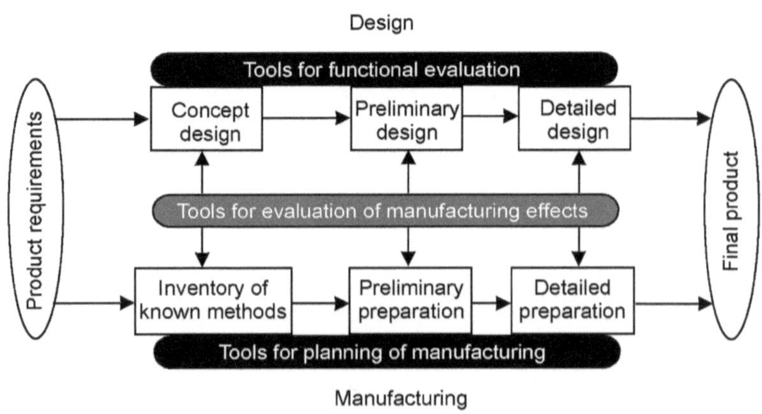

13.1 Key activities during the product and process development.

projects. Thus it has been important to apply working methods where these tasks are carried out ahead of the actual use of the models in development projects, as the time schedule is very tight at this stage.

13.4 Examples

This section gives examples of development projects at Volvo Aero where simulations of welding have played an important part.

13.4.1 ECAPS

ECAPS is a satellite rocket for orbit adjustment made of rhenium (Fig. 13.2). Several processes are used in manufacturing the rocket. One of them is EB welding. Welding trials were carried out during the product development. They showed that a large region was melting and damaging the rocket. Different welding methods were evaluated by simulations in order to solve this problem. The simulations showed that EB welding was impossible given the design requirement and welding as a joining technique in this project was abandoned. This project demonstrated that costly and time-consuming physical trials could have been avoided by using the simulations earlier.

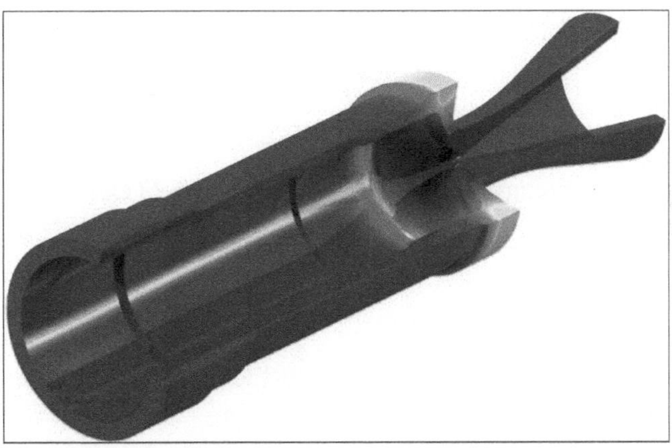

13.2 The ECAPS rocket. Temperature distribution during welding to large meted zone, grey area, is observed in the product.

13.4.2 Low-pressure turbine case

One of the rear components in an aero-engine is the low-pressure turbine case (LPT-case) shown in Fig. 13.3. LPT-Cases consist of several rings that are EB welded together. However, welding caused cracks that were found when the welded component was inspected.

13.3 Low-pressure turbine case.

Welding simulations were used in order to understand the problem and identify a possible solution. It was found that the positions of the weld start and weld stop had an influence on the risk of cracking. Moreover, it was shown that if material close to the weld zone was removed, a more beneficial stress pattern was created as a result of welding (see Fig. 13.4). This project showed how simulations can lead to an increased understanding of the problem and make it possible to evaluate several possible solutions at a much lower cost than by the use of physical trials.

13.4.3 PW2000 – turbine exhaust case

Volvo Aero is manufacturing the turbine exhaust case (TEC) of the PW2000 aircraft engine (Fig. 13.5). The TEC is a structural component in the rear part of

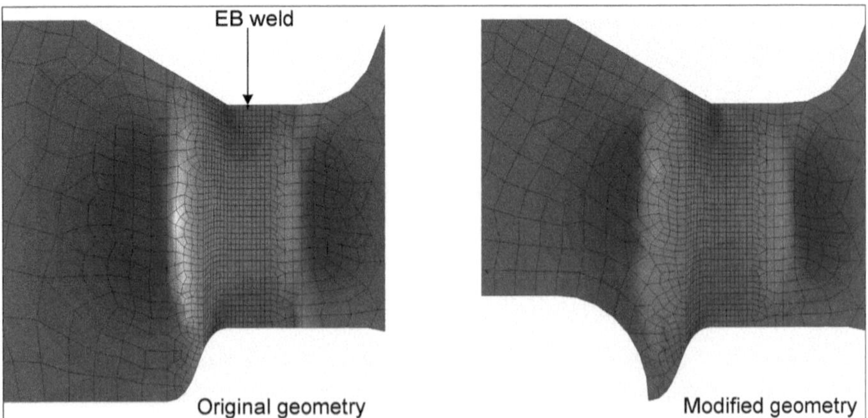

13.4 Residual hoop stress pattern after EB welding.

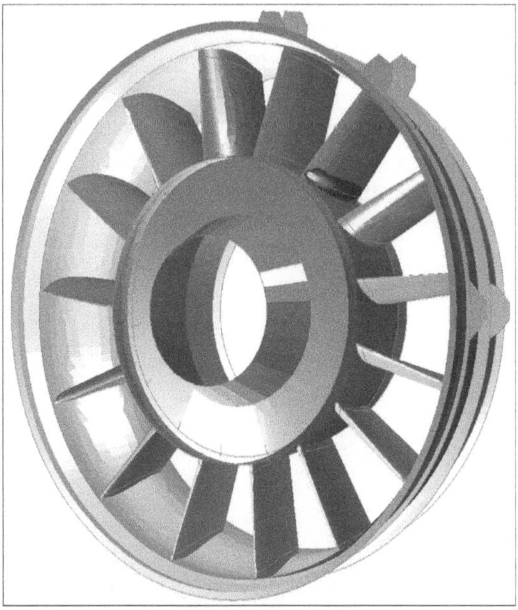

13.5 Welding simulation to investigate the influence of welding sequence on the PW2000 turbine exhaust case.

an aircraft engine. Pratt & Whitney's PW2000 engine covers a thrust range from 37 000–43 000 lbs (16 650–19 350 kg). The PW2000-TEC is manufactured by fabrication and about 200 welds are needed to manufacture the product. There have been issues with geometrical tolerances in production. Several welding sequences were investigated by use of welding simulations in order to meet these tolerances. These simulations showed that the residual stresses could be reduced by an appropriate welding sequence. Moreover, a pre-deformation was given to the parts before welding, leading to reduced deformations after welding that complied with the geometric tolerances. The amount of pre-deformation was also determined by the welding simulation.

13.4.4 GEnx – turbine rear frame

The GEnx-engine family is General Electric's next generation aircraft engine with a thrust range of 53 000–75 000 lbs (23 850–33 750 kg). Volvo Aero is responsible for the design and manufacture of the turbine rear frame (TRF) of the GEnx aero-engine. The TRF is a structural component in the rear part of an engine. The chosen manufacturing technique of the TRF is fabrication and the number of welding operations required to manufacture the product exceeds one hundred. Welding simulation has been used in order to fulfil the geometrical tolerances after the welding of the TRF. A close cooperation between welding analysts, welders, welding engineers and fixture design engineers, and the simulation of many welding sequences, resulted in an optimal welding sequence which met the geometrical tolerances required.

The use of welding simulations enabled the investigation of welding sequences that previously were too complex and costly to carry out by physical trials. They decreased the product development time and cost, and an optimal welding sequence was obtained that fulfilled the tolerances, i.e. 'right first time'.

14

Summary and conclusion

The development from simple laboratory set-up to large industrial application in CWM has been noted already in the review by Lindgren [86]. This is even more obvious now when comparing the applications in Chapter 13 with some of the initial publications in CWM. It is not only the size of the models (Fig. 1.1) that has increased; they have also accommodated better material models and other submodels (see Chapter 9). The trend in the development in CWM goes in two directions: larger and smaller scales.

Development of simulation tools in combination with the improvements in computer hardware will enable simulation of even larger models. However, there are issues that must be dealt with by the research community in CWM. One of them is the need for 'integrated modelling'. This concept, introduced by Radaj [424], denotes cases where CWM results are also used in subsequent analyses for estimating risk for buckling, fatigue, etc.

Welding is only one of the manufacturing steps in production. It is necessary to simulate a chain of manufacturing processes in order to estimate their cumulative effect on the component. Heat treatment is often used in conjunction with welding and therefore this has been the first process of interest. There are many publications in the field of heat treatment simulations [441] and there are several similarities with welding simulations facilitating a two-step simulation, welding and post-weld heat treatment. Less common is the combination of forming and welding simulations. Forming simulations are probably the most common type of simulations in industry. Therefore, it is not difficult to include them. This kind of simulation requires material model(s) for the varying processes, adaptation of meshes together with data transfer between each step.

The development in the smaller scale is concerned with material modelling. The coupling of material models with microstructure models is of special interest in multipass and temper bead welding where a range of microstructures with varying properties exists. There are models that can accurately represent the near-weld pool behaviour for prediction of hot cracking and coupling with diffusion of, for example, hydrogen. Simulations within the field of CWM will also benefit from weld process models, such as the work by Sudnik [16], that can predict the heat generation.

15
References

1. Lindgren, L.-E., A. Carlestam, and M. Jonsson, Computational model of flame-cutting. *ASME Journal of Engineering Materials and Technology*, 1993. **115**: 440–445.
2. Adams, V. and A. Askenazi, *Building Better Products with Finite Element Analysis*. 1999: OnWord Press.
3. Lindgren, L.-E., H. Runnemalm, and M.O. Nasstrom, Simulation of multipass welding of a thick plate. *International Journal for Numerical Methods in Engineering*, 1999. **44**(9): 1301–1316.
4. Zacharia, T., *et al.*, Modeling of fundamental phenomena in welds. *Modelling and Simulation in Materials Science and Engineering*, 1995. **3**: 265–288.
5. Akhlagi, M. and J. Goldak, *Computational Welding Mechanics*. 2005: Springer.
6. Radaj, D., Schweissprozess-simulation. Grundlagen und Anwendungen (in German). *Fachbuche Schweisstechnik*, 1999. **141**, p. 193.
7. Radaj, D., *Welding residual stresses and distortion. Calculation and measurement.* 2003: DVS-Verlag.
8. Ramirez, M., G. Trapaga, and J. McKelliget, A comparison between two different numerical formulations of welding arc simulation. *Modelling and Simulation in Materials Science and Engineering*, 2003. **11**(4): 675–695.
9. Zhu, P. Computer simulation of gas metal welding arcs. In *5th International Conference on Trends in Welding Research*. 1998. Pine Mountain, Georgia: American Welding Society.
10. DebRoy, T., Mathematical modelling of fluid flow and heat transfer in fusion welding. In *Mathematical Modelling of Weld Phenomena 5*. 1999. Seggau, Austria: IOM Communications.
11. Pavlyk, V. and U. Dilthey, Simulation of weld solidification microstructure and its coupling to the macroscopic heat and fluid flow modelling. Vol. 12. 2004. S33. http://www.iop.org/EJ/abstract/0965-0393/12/1/503
12. Do-Quang, M. and G. Amberg, Modelling of time-dependent 3D weld pool flow. In *Mathematical Modelling of Weld Phenomena 7*. 2003. Graz, Austria.
13. Winkler, C., *et al.*, The effect of surfactant redistribution on the weld pool shape during GTA-welding. *Science and Technology of Welding and Joining*, 2000. **5**: 1–13.
14. Pavlyk, V. and U. Dilthey, A numerical and experimental study of fluid flow and heat transfer in stationary GTA weld pools. In *Mathematical Modelling of Weld Phenomena 5*. 1999. Seggau, Austria: IOM Communications.
15. Sudnik, W., D. Radaj, and W. Erofeew, Computerized simulation of laser beam

weld formation comprising joint gaps. *Journal of Physics D: Applied Physics*, 1998. **31**(24): 3475.
16. Sudnik, W., *et al.*, Numerical simulation of weld pool geometry in laser beam welding. *Journal of Physics D: Applied Physics*, 2000. **33**(6): 662.
17. Lundbäck, A. and H. Runnemalm, Validation of three dimensional finite element model in electron beam welding of Inconel 718. *Science and Technology of Welding and Joining*, 2005. **10**(6): 717–724.
18. Sheng, I. and Y. Chen, Modeling welding by surface heating. *ASME Journal of Engineering Materials and Technology*, 1992. **114**(4): 439–449.
19. Chen, Y. and I. Sheng, On the solid-fluid transition zone in welding analysis. In *Heat and Mass Transfer in Solidification Processing*. 1991: ASME, pp. 21–33.
20. Felippa, C., K. Park, and C. Farhat, Partitioned analysis of coupled mechanical systems. *Computer Methods in Applied Mechanics and Engineering*, 2001. **190**: 3247–3270.
21. Felippa, C., K. Park, and C. Farhat, *Partitioned Analysis of Coupled Mechanical Systems*. 1999: Center for Aerospace Structures, University of Colorado, p. 31.
22. Simo, J.C. and C. Miehe, Associative coupled thermoplasticity at finite strains: formulation, numerical analysis and implementation. *Computer Methods in Applied Mechanics and Engineering*, 1992. **98**(1): 41–104.
23. Armero, F. and J.C. Simo, A new unconditionally stable fractional step method for nonlinear coupled thermomechanical problems. *International Journal for Numerical Methods in Engineering*, 1992. **35**: 737–766.
24. Lewis, R., P. Nithiarasu, and K. Seetharamu, *Fundamentals of the Finite Element Method for Heat Transfer and Fluid Flow*. 2004: John Wiley & Sons.
25. Hsu, T.-R., *The Finite Element Method in Thermomechanics*. 1986: Allen & Unwin.
26. Ziegler, H., *An Introduction to Thermomechanics*. 2nd edm, Series in Applied Mathematics and Mechanics. 1983: North-Holland.
27. Sluzalec, A., *Introduction to Nonlinear Thermomechanics*. 1991: Springer-Verlag.
28. Fung, Y., *Foundations of Solid Mechanics*. 1965: Prentice-Hall.
29. Boley, B. and J. Weiner, *Theory of Thermal Stresses*. 1960: John Wiley & Sons.
30. Carslaw, H. and J. Jaeger, *Conduction of Heat in Solids*, 2nd edn, 1959: Clarendon Press.
31. Huang, H.-C. and A. Usmani, *Finite Element Analysis for Heat Transfer*. 1994: Springer Verlag.
32. Bathe, K.-J., *Finite Element Procedures*. 1996: Prentice-Hall.
33. Cook, R., D. Malkus, and M. Plesha, *Concepts and Applications of Finite Element Analysis*. 1989: John Wiley & Sons.
34. Belytschko, T., W.K. Liu, and B. Moran, *Nonlinear Finite Elements for Continua and Structures*. 2000: John Wiley & Sons.
35. Belytschko, T. and T. Hughers, *Nonlinear Finite Element Analysis – A Short Course*. 2003: Paris.
36. Wang, J., *et al.*, Prediction of welding deformations by FEM based on inherent strains. In *8th International Conference of Casting, Welding and Advanced Solidification Processes*. 1998. San Diego, USA.
37. Ronda, J. and G.J. Oliver, Consistent thermo-mechano-metallurgical model of welded steel with unified approach to derivation of phase evolution laws and transformation-induced plasticity. *Computer Methods in Applied Mechanics and Engineering*, 2000. **189**(2): 361–418.
38. Armero, F. and J.C. Simo, A priori stability estimates and unconditionally stable product formula algorithms for nonlinear coupled thermoplasticity. *International*

Journal of Plasticity, 1993. **9**(6): 749–782.
39. Karlsson, L. and L.-E. Lindgren. Combined heat and stress-strain calculations. In *Modeling of Casting, Welding and Advanced Solidification Processes V*. 1990. Davos, Switzerland: The Minerals, Metals & Materials Society.
40. Goldak, J., M. Gu, and E. Hughes, Steady state thermal analysis of welds with filler metal addition. *Canadian Metallurgical Quarterly*, 1993. **32**(1): 49–55.
41. Balagangadhar, D., *et al.*, A displacement-based reference frame formulation for steady-state thermo-elasto-plastic material processes. *International Journal of Solids and Structures*, 1999. **36**(16): 2397–2416.
42. Rajadhyaksha, S. and P. Michaleris, Optimization of thermal processes using an Eulerian formulation and application in laser hardening. *International Journal for Numerical Methods in Engineering*, 2000. **47**: 1807–1823.
43. Shanghvi, J. and P. Michaleris, Thermo-elasto-plastic finite element analysis of quasi-state processes in Eulerian reference frames. *International Journal for Numerical Methods in Engineering*, 2002. **53**: 1533–1556.
44. Paul, S., P. Michaleris, and J. Shangvi, Optimization of thermo-elasto-plastic processes using Eulerian sensitivity analysis. *International Journal for Numerical Methods in Engineering*, 2003. **56**: 1125–1150.
45. Zhang, L. and P. Michaleris, Investigation of Lagrangian and Eulerian finite element methods for modeling the laser forming process. *Finite Elements in Analysis and Design*, 2004. **40**: 383–405.
46. Bergheau, J., D. Pont, and J. Leblond, Three-dimensional simulation of a laser surface treatment through steady state computation in the heat source's comoving frame. In *IUTAM Symposium on the Mechanical Effects of Welding*. 1992. Luleå, Sweden: Springer Verlag.
47. Gu, M., *et al.*, Modelling the evolution of microstructure in the heat affected zone of steady state welds. *Canadian Metallurgical Quarterly*, 1993. **32**: 351–363.
48. Gu, M. and J. Goldak, Steady-state formulation for stress and distortion of welds. *ASME Journal of Engineering for Industry*, 1994. **116**: 467–474.
49. Goldak, J. and M. Gu, Computational weld mechanics of the steady state. In *Mathematical Modelling of Weld Phenomena 2*. 1995. The Institute of Materials.
50. Atteridge, D., *et al.*, 3D Eulerian modeling of electroslag welding. In *5th International Conference on Trends in Welding Research*. 1998. Pine Mountain, Georgia, USA.
51. Goldak, J. and M. Akhlagi, *Computational Welding Mechanics*. 2005: Springer.
52. Rosenthal, D., Mathematical theory of heat distribution during welding and cutting. *Welding Journal*, 1941. **20**(5): 220s–234s.
53. Rosenthal, D., The theory of moving sources of heat and its application to metal treatments. *Transactions of ASME*, 1946. **68**(Nov): 849–866.
54. Codina, R., *Finite Element Formulation for the Numerical Solution of the Convection–Diffusion Equation*, Vol. 14. 1993: CIMNE.
55. Hughes, T., Recent progress in the development and understanding of SUPG methods with a special reference to the compressible Euler and Navier–Stokes equations. *International Journal for Numerical Methods in Fluids*, 1987. **7**(1): 1261–1275..
56. Onate, E., J. Miquel, and G. Hauke, Stabilized formulation for the advection-diffusion-absorption equation using finite calculus and linear finite elements. *Computer Methods in Applied Mechanics and Engineering*, 2006. **195**: 3926–3946.
57. Levitas, V., *et al.*, Numerical modelling of martensitic growth in an elastoplastic material. *Philosophical Magazine A*, 2002. **82**(3): 429–462.

58. Granjon, H., *Fundamentals of Welding Metallurgy*. 1991: Woodhead Publishing.
59. Easterling, K., *Introduction to the Physical Metallurgy of Welding*. 1983: Butterworths.
60. Grøng, Ø., *Metallurgical Modelling of Welding*. 1994: The Institute of Materials.
61. Garcia de Andrés, C., *et al.*, Application of dilatometric analysis to the study of solid-solid phase transformation in steels. *Materials Characterization*, 2002. **48**: 101–111.
62. Jäniche, W., *et al.*, eds, *Steel: A Handbook for Materials Research and Engineering. Volume 1: Fundamentals*. 1992: Verlag Stahleisen.
63. Fuglsang Andersen, L., *Residual Stresses and Deformations in Steel Structures*. 2000: Department of Naval Architecture and Offshore Engineering, Technical University of Denmark.
64. Leblond, J., *et al.*, Metallurgical and mechanical consequences of phase transformations in numerical simulations of welding processes. In *Modeling in Welding, Hot Powder Forming, and Casting*, L. Karlsson, Editor. 1997: ASM International, pp. 61–89.
65. Börjesson, L. and L.-E. Lindgren, Thermal, metallurgical and mechanical models for simulation of multipass welding. *ASME Journal of Engineering Materials and Technology*, 2001. **123**(Jan): 106–111.
66. Satoh, K., Transient thermal stresses of weld heat-affected zone by both-ends-fixed bar analogy. *Transactions of the Japan Welding Society*, 1972. **3**(1): 125–134.
67. Okerblom, N., *The Calculation of Deformations of Welded Metal Structures. A Translation from the Russian by LC Ronson.* 1958: Her Majesty's Stationery Office.
68. Bhadeshia, H., *et al.*, Cancellation of residual stress using transformation plasticity. In *8th International Seminar Numerical Analysis of Weldability*. 2006. Seggau, Austria.
69. Vincent, Y., J.-F. Jullien, and P. Gilles, Thermo-mechanical consequences of phase transformations in the heat-affected zone using a cyclic uniaxial test. *International Journal of Solids and Structures*, 2005. **42**(14): 4077–4098.
70. Pilipenko, A., *Computer Simulation of Residual Stress and Distortion of Thick Plates in Multielectrode Submerged Arc Welding – Their Mitigation Techniques*. 2001: Department of Machine Design and Materials Technology, Norwegian University of Science and Technology.
71. Jonsson, M., L. Karlsson, and L.-E. Lindgren, Deformations and stresses in butt-welding of large plates. In *Numerical Methods in Heat Transfer*, R. Lewis, Editor. 1985: Wiley, pp. 35–58.
72. Berglund, D. and H. Alberg, A two-stage approach for validation of welding- and heat treatment models in product development. *Science and Technology of Welding and Joining*, 2005. **10**(6): 653–665.
73. Stone, H., S. Roberts, and R. Reed, A process model for the distortion induced by the electron-beam welding of a nickel-based superalloy. *Metallurgical and Materials Transactions A*, 2000. **31A**(Sep): 2261–2273.
74. Dhingra, A.K. and C.L. Murphy, Numerical simulation of welding-induced distortion in thin-walled structures. *Science and Technology of Welding and Joining*, 2005. **10**(5): 528–536.
75. Karlsson, R. and B. Josefson, Three-dimensional finite element analysis of temperatures and stresses in a single-pass butt-welded pipe. *ASME Journal of Pressure Vessel Technology*, 1990. **112**: 76–84.
76. Josefson, L. and C. Karlsson, Transformation plasticity effects in residual stresses

in a butt-welded pipe. *ASME Journal of Pressure Vessel Technology*, 1992. **114**: 376–378.
77. Murthy, Y.V.L.N., G.V. Rao, and P.K. Iyer, Numerical simulation of welding and quenching processes using transient thermal and thermo-elasto-plastic formulations. *Computers and Structures*, 1996. **60**(1): 131–154.
78. Troive, L., L.-E. Lindgren, and M. Jonsson, Axial collapse load of a girth butt-welded pipe. In *Thermal Stresses '95*. 1995. Hamamatsu, Japan.
79. Comini, G., *et al.*, Finite element solution of non-linear heat conduction problems with special reference to phase change. *International Journal for Numerical Methods in Engineering*, 1974. **8**: 613–624.
80. Chester, M., Second sound in solids. *Physical Review*, 1963. **131**(5): 2013–2015.
81. Ali, Y. and L. Zhang, Relativistic heat conduction. *International Journal of Heat and Mass Transfer*, 2005. **48**(12): 2397–2406.
82. Pham, Q., Comparison of general purpose finite element methods for the Stefan problem. *Numerical Heat Transfer Part B – Fundamentals*, 1995. **27**(4): 417–435.
83. Lewis, R., *et al.*, *The Finite Element Method in Heat Transfer Analysis*. 1996: John Wiley & Sons.
84. Belytschko, T., An overview of semidiscretization and time procedures. In *Computational Methods in Mechanics*, T. Belytschko and K.-J. Bathe, Editors. 1983: North-Holland, pp. 3–65.
85. Hughes, T., Analysis of transient algorithms with particular reference to stability behavior. In *Computational Methods for Transient Analysis*, T. Belytschko and T. Hughes, Editors. 1983: North-Holland, pp. 68–155.
86. Lindgren, L.-E., Finite element modelling and simulation of welding, Part 1 Increased complexity. *Journal of Thermal Stresses*, 2001. **24**: 141–192.
87. Bergman, G. and M. Oldenburg, A finite element model for thermomechanical analysis of sheet metal forming. *International Journal for Numerical Methods in Engineering*, 2004. **59**(9): 1167–1186.
88. Andersson, B.A.B., Thermal stresses in a submerged-arc welded joint considering phase transformations. *ASME Journal of Engineering Materials & Technology*, 1978. **100**: 356–362.
89. Karlsson, L., Thermal stresses in welding. In *Thermal Stresses*, R.B. Hetnarski, Editor. 1986: Elsevier Science Publishers, pp. 300–389.
90. Rolph, W.D. and K.-J. Bathe, An efficient algorithm for analysis of non linear heat transfer with phase changes. *International Journal of Numerical Methods in Engineering*, 1982. **18**: 119–134.
91. Banaszek, J., Comparison of control volume and Galerkin finite element methods for diffusion-type problems. *Numerical Heat Transfer*, 1989. **16**: 59–78.
92. Crisfield, M.A., *Non-linear Finite Element Analysis of Solids and Structures, Volume 1: Essentials*. 1991: John Wiley & Sons.
93. Crisfield, M.A., *Non-Linear Finite Element Analysis of Solids and Structures, Vol. 2 Advanced Topics*. 1997: John Wiley & Sons.
94. Bonet, J. and R.D. Wood, *Nonlinear Continuum Mechanics for Finite Element Analysis*. 1997: Cambridge University Press.
95. Simo, J.C. and T.J.R. Hughes, *Computational Inelasticity*, J.E. Marsden, L. Sirovich, and S. Wiggins, Editors. 1997: Springer-Verlag.
96. Belytschko, T. and T. Hughes, eds, *Computational Methods for Transient Analysis. Computational Methods in Mechanics. Vol. 1*. 1983: North-Holland.
97. Malvern, L.E., *Introduction to the Mechanics of a Continuous Medium*. Engineering

of the Physical Sciences Series, J. Reswick and W. Rohsenow, Editors. 1969: Prentice-Hall.

98. Chung, T., *Applied Continuum Mechanics*. 1996: Cambridge University Press.
99. Simo, J. and R. Taylor, Consistent tangent operators for rate-independent elastoplasticity. *Computer Methods in Applied Mechanics and Engineering*, 1985. **48**: 101–118.
100. Ponthot, J.P., Unified stress update algorithms for the numerical simulation of large deformation elasto-plastic and elasto-viscoplastic processes. *International Journal of Plasticity*, 2002. **18**(1): 91–126.
101. Hughes, T., Numerical implementation of constitutive models: rate-independent deviatoric plasticity. In *Theoretical Foundation for Large Scale Computation of Nonlinear Material Behaviour*. 1984. Northwestern University, Evanston, Illinois: Nijhoff.
102. Simo, J. and K. Pister, Remarks on rate constitutive constitutive equations for finite deformation problems: computational implications. *Computer Methods in Applied Mechanics and Engineering*, 1985. **46**: 201–215.
103. Arif, A., T. Pervez, and M. Mughal, Performance of a finite element procedure for hyperelastic-viscoplastic large deformation problems. *Finite Elements in Analysis and Design*, 2000. **34**: 89–112.
104. Simo, J.C., A framework for finite strain elastoplasticity based on maximum plastic dissipation and the multiplicative decomposition: Part I. Continuum formulation. *Computer Methods in Applied Mechanics and Engineering*, 1988. **66**: 199–219.
105. Simo, J.C., A framework for finite strain elastoplasticity based on maximum plastic dissipation and the multiplicative decomposition: Part II. Computational aspects. *Computer Methods in Applied Mechanics and Engineering*, 1988. **68**: 1–31.
106. Hughes, T. and J. Winget, Finite rotation effects in numerical integration of rate constitutive equations arising in large-deformation analysis. *International Journal for Numerical Methods in Engineering*, 1980. **15**(9): 1862–1867.
107. Johnson, G. and D. Bammann, A discussion of stress rates in finite deformation problems. *International Journal of Solids and Structures*, 1984. **22**(8): 725–737.
108. Szabó, L. and M. Balla, Comparison of some stress rates. *International Journal of Solids and Structures*, 1989. **25**(3): 279–297.
109. Moss, C., On instabilities in large deformation simple shear loading. *Computer Methods in Applied Mechanics and Engineering*, 1984. **46**: 329–338.
110. Dienes, J., On the analysis of rotation and stress rate in deforming bodies. *Acta Mechanica*, 1979. **32**: 217–232.
111. Marin, E. and D. McDowell, A semi-implicit integration scheme for rate-dependent and rate-independent plasticity. *Computers and Structures*, 1997. **63**(3): 579–600.
112. Chaboche, J.L. and G. Cailletaud, Integration methods for complex plastic constitutive equations. *Computer Methods in Applied Mechanics and Engineering*, 1996. **133**(1–2): 125–155.
113. Saleeb, A., T. Wilt, and W. Li, Robust integration schemes for generalized viscoplasticity with internal-state variables. *Computers and Structures*, 2000. **74**: 601–628.
114. Ramm, E. and A. Matzenmiller, Consistent linearization in elasto-plastic shell analysis. *Engineering Computations*, 1988. **5**(Dec): 289–299.
115. Perzyna, P. and R. Pecherski, Modified theory of viscoplasticity: physical foundations and identification of material functions for advanced strains. *Archives of Mechanics*, 1983. **35**(3): 423–436.
116. Miller, A.K., The MATMOD equations. In *Unified Constitutive Equations for*

Creep and Plasticity, A.K. Miller, Editor. 1987: Elsevier Applied Science Publishers, pp. 139–219.

117. Kassner, M.E. and M.-T. Pérez-Prado, Five-power-law creep in single phase metals and alloys. *Progress in Materials Science*, 2000. **45**: 1–102.
118. Frost, H.J. and M.F. Ashby, Deformation-mechanism maps for pure iron, two austenitic stainless steels and a low-alloy ferritic steel. In *Fundamental Aspects of Structural Alloy Design*, R.I. Jaffee and B.A. Wilcox, Editors. 1977: Plenum Press, pp. 26–65.
119. Frost, H.J. and M.F. Ashby, *Deformation-mechanism Maps – The Plasticity and Creep of Metals and Ceramics*. 1982: Pergamon Press.
120. Miller, A.K., ed., *Unified Constitutive Equations for Creep and Plasticity*. 1987: Elsevier Applied Science Publishers Ltd.
121. Raboin, P., *A Deformation-mechanism Material Model for Nike3D*. 1993: Lawrence Livermore National Laboratory.
122. de Souza Neto, E., A simple robust numerical integration algorithm for a power-law visco-plastic model under both high and low rate-sensitivity. *International Journal for Numerical Methods in Engineering*, 2004. **20**: 1–17.
123. Oddy, A., J. McDill, and J. Goldak, Consistent strain fields in 3D finite element analysis of welds. *ASME Journal of Pressure Vessel Technology*, 1990. **112**(3): 309–311.
124. McDill, J. and A. Oddy, A non-conforming eight to 26-node hexahedron for three-dimensional thermal-elastoplastic finite element analysis. *Computers and Structures*, 1995. **54**(2): 183–189.
125. Friedman, E., Thermomechanical analysis of the welding process using the finite element method. *ASME Journal of Pressure Vessel Technology*, 1975. **97**(3): 206–213.
126. Friedman, E. Numerical simulation of the gas tungsten-arc welding process. In *ASME Winter Annual Meeting*. 1977. Atlanta, USA.
127. Hibbitt, H. and P. Marcal, A numerical thermo-mechanical model for the welding and subsequent loading of a fabricated structure. *Computers and Structures*, 1973. **3**: 1145–1174.
128. Benzley, S., *et al.* A comparison of all hexagonal and all tetrahedral finite element meshes for elastic and elasto-plastic analysis. In *Proceedings of 14th Annual International Meshing Roundtable*. 1995. Albuquerque, USA.
129. Cifuentes, A. and A. Kalbag, A performance study of tetrahedral and hexahedral elements in 3-D finite element structural analysis. *Finite Element in Analysis and Design*, 1992. **12**: 313–331.
130. Owen, S., A survey of unstructured mesh generation technology. 2006 [cited 2006 4 Nov]; available from: http://www.andrew.cmu.edu/user/sowen/survey/index.html.
131. Owen, S., Meshing software survey. 2006 [cited 2006 4 Nov]; available from: http://www.andrew.cmu.edu/user/sowen/softsurv.html.
132. Hyun, S., *Adaptive Remeshing and Simulating a Chain of Manufacturing Processes*. 2002: Department of Mechanical Engineering, Luleå University of Technology.
133. McDill, J.M., *et al.*, Isoparametric quadrilaterals and hexahedrons for mesh-grading algorithms. *Communication in Applied Numerical Methods*, 1987. **3**: 155–163.
134. McDill, J., K. Runnemalm, and A. Oddy, An 8- to 16-node solid graded shell element for far-field applications in 3-D thermal-mechanical FEA. *Mathematical Modelling and Scientific Computing*, 2001. **13**(3): 177–192.
135. Tan, X. and L. Vu-Quoc, Efficient and accurate multilayer solid-shell element: non-

linear materials at finite strain. *International Journal for Numerical Methods in Engineering*, 2005. **63**(15): 2124–2170.

136. Lindgren, L.-E., *et al.*, Automatic remeshing for three-dimensional finite element simulation of welding. *Computer Methods in Applied Mechanics and Engineering*, 1997. **147**(3–4): 401–409.
137. Stadthaus, M., V. Michailov, and H. Wohlfahrt, Simulation of weld shape, microstructure, distortion and residual stresses for laser beam welding. In *Mathematical Modelling of Weld Phenomena 6*. 2001. Graz, Austria: Institute of Materials, Minerals and Mining.
138. Lenz, B. and F. Rick, Lösungsansatze zur Simulation des Laserstrahlschweissens mit thermomechanischen und mechanischen Analysen. In *MARC Benutzertreffen*. 1999: Munchen.
139. Runnemalm, H. and S. Hyun, Three-dimensional welding analysis using an adaptive mesh scheme. *Computer Methods in Applied Mechanics and Engineering*, 2000. **189**: 515–523.
140. Hemmer, H., S. Klokkehaug, and O. Grong, Process modelling applied to electron beam welding of duplex stainless steels. In *Mathematical Modelling of Weld Phenomena 6*. 2001. Graz, Austria: The Institute of Materials, Minerals and Mining.
141. Rieger, A. and P. Wriggers, Adaptive methods for thermomechanical coupled contact problems. *International Journal for Numerical Methods in Engineering*, 2004. **59**: 871–894.
142. Dow, J., *A Unified Approach to the Finite Element Method and Error Analysis Procedures*. 1999: Academic Press.
143. Barthold, F.-J., M. Schmidt, and E. Stein, Error indicators and mesh refinements for finite-element-computations of elastoplastic deformations. *Computational Mechanics*, 1998. **22**(3): 225–238.
144. Zhu, J.Z., A posteriori error estimation – the relationship between different procedures. *Computer Methods in Applied Mechanics and Engineering*, 1997. **150**: 411f.
145. Zienkiewicz, O.C. and J.Z. Zhu, A simple error estimator and adaptive procedure for practical engineering analysis. *International Journal for Numerical Methods in Engineering*, 1987. **24**: 337–357.
146. Zienkiewicz, O.C. and J.Z. Zhu, The superconvergent patch recovery and *a posteriori* error estimates, Part I: The recovery technique. *International Journal for Numerical Methods in Engineering*, 1992. **33**: 1331–1364.
147. Zienkiewicz, O.C., B. Boroomand, and J.Z. Zhu, Recovery procedures in error estimation and adaptivity Part I: Adaptivity in linear problems. *Computer Methods in Applied Mechanics and Engineering*, 1999. **176**(1–4): 111–125.
148. Lindgren, L.-E. and E. Hedblom, Modelling of addition of filler material in large deformation analysis of multipass welding. *Communications in Numerical Methods in Engineering*, 2001. **17**(9): 647–657.
149. Huerta, A., *et al.*, Adaptive finite element strategies based on error assessment. *International Journal for Numerical Methods in Engineering*, 1999. **46**: 1803–1818.
150. Boroomand, B. and O.C. Zienkiewicz, Recovery procedures in error estimation and adaptivity. Part II: Adaptivity in nonlinear problems of elasto-plasticity behaviour. *Computer Methods in Applied Mechanics and Engineering*, 1999. **176**(1–4): 127–146.
151. Hyun, S. and L.-E. Lindgren, Smoothing and adaptive remeshing schemes for graded element. *Communications in Numerical Methods in Engineering*, 2001. **17**: 1–17.

152. Oddy, A., *et al.*, A distortion metric for iso-parametric finite elements. *Canadian Society of Mechanical Engineers*, 1988. **12**(4): 213–217.
153. Andersen, L.F., *Residual Stresses and Deformations in Steel Structures*. 2000: Department of Naval Architecture and Offshore Engineering, Technical University of Denmark.
154. Souloumiac, B., F. Boitout, and J. Bergheau, A new local-global approach for the modelling of welded steel component distortions. In *Mathematical Modelling of Weld Phenomena 6*. 2001. Graz, Austria: Institute of Materials, Minerals and Mining.
155. Faure, F., *et al.*, Prediction of distortions of large thins structures during welding using shell elements and multiscale approaches. In *Mathematical Modelling of Weld Phenomena 7*. 2003. Graz, Austria.
156. Ueda, Y., Predicting and Measuring methods of two- and three-dimensional welding residual stresses by using inherent strain as a parameter. In *Modeling in Welding, Hot Powder Forming and Casting*, L. Karlsson, Editor. 1997: ASM International, pp. 91–113.
157. Murakawa, H., Y. Luo, and Y. Ueda, Inherent strain as an interface between computational welding mechanics and its industrial application. In *Mathematical Modelling of Weld Phenomena 4*. 1998. Graz, Austria: Institute of Materials.
158. Ericsson, M., D. Berglund, and P. Nylén, Three dimensional simulation of robot path, heat transfer and residual stresses of a TIG-welded part with complex geometry. In *6th International Conference on Trends in Welding Research*. 2002. Pine Mountain, Georgia: ASM International.
159. Iwaki, T. and K. Masubuchi, Thermo-elasto-plastic analysis of orthotropic plate by the finite element method (in Japanese). *Journal of the Society of Naval Architects of Japan*, 1971. **130**: 193–202.
160. Ueda, Y. and T. Yamakawa, Thermal stress analysis of metals with temperature dependent mechanical properties. In *International Conference on Mechanical Behaviour of Materials*. 1971. Kyoto.
161. Ueda, Y. and T. Yamakawa, Analysis of thermal elastic-plastic stress and strain during welding by finite element method. *Transactions of the Japan Welding Research Institute*, 1971. **2**(2): 90–100.
162. Fujita, Y. and T. Nomoto, Studies on thermal stresses in welding with special reference to weld cracking. In *International Conference on Mechanical Behaviour of Materials*. 1971. Tokyo.
163. Fujita, Y., *et al.*, *Welding Stresses with Special Reference to Cracking*. 1972: International Institute of Welding.
164. Lindgren, L.-E. and L. Karlsson, Deformations and stresses in welding of shell structures. *International Journal for Numerical Methods in Engineering*, 1988. **2**: 155–164.
165. McDill, J., *et al.*, Finite element analysis of weld distortion in carbon and stainless steels. *Journal of Strain Analysis*, 1990. **25**(1): 51–53.
166. Goldak, J., *et al.* Computational weld mechanics. In *AGARD Workshop – Structures and Materials 61st Panel Meeting*. 1985. Oberammergau, Germany.
167. McDill, JMJ, A. Oddy, and J. Goldak, Comparing 2-D plane strain and 3-D analyses of residual stresses in welds. In *The 3rd International Conference on Trends in Welding Research*. 1992. Gatlinburg, Tennessee.
168. Ma, N.-X., *et al.*, FEM analysis of 3-D welding residual stresses and angular distortion in T-type fillet welds. *Transactions of the Japan Welding Institute*, 1995. **24**(2): 115–122.

169. Papazoglou, V.J. and K. Masubuchi, Numerical analysis of thermal stresses during welding including phase transformation effects. *ASME Journal of Pressure Vessel Technology*, 1982. **104**: 198–203.
170. Leung, C. and R. Pick, The use of generalized plane strain elements in the prediction of residual stresses in welded flat plates. In *Computer Modelling of Fabrication Processes and Constitutive Behaviour of Metals*. 1986. Ottawa: Canada Centre for Mineral and Energy Technology.
171. Troive, L., *et al.*, Finite element simulations of the bending of a flat plate to U-shaped beam cross-section and the welding to rectangular hollow cross-section and neutron diffraction determination of residual stresses. In *The Second International Conference on Recent Trends in Welding Research*. 1989. Gatlinburg, USA.
172. Mok, D. and R. Pick, Finite element study of residual stresses in a plate T-joint fatigue specimen. *Journal of Mechanical and Engineering Science*, 1990. **204**(C2): 127–134.
173. Wikander, L., *et al.*, Finite element simulation and measurement of welding residual stresses. *Modelling and Simulation in Materials Science and Engineering*, 1994. **2**(4): 845–864.
174. Roelens, J., Determination of residual stresses in submerged arc multi-pass welds by means of numerical simulation and comparisons with experimental measurements. In *Mathematical Modelling of Weld Phenomena 2*. 1995. Graz, Austria: The Institute of Materials.
175. Michaleris, P. and X. Sun, Finite element analysis of thermal tensioning techniques mitigating weld buckling distortion. *Welding Journal*, 1997. **76**(11): 451s–457s.
176. Michaleris, P., Z. Feng, and G. Campbell, Evaluation of 2D and 3D FEA models for predicting residual stress and distortion. *Approximate Methods in the Design and Analysis of Pressure Vessels and Piping Components*, American Society of Mechanical Engineers, Pressure Vessels and Piping Division (Publication) PVP, 1997. **347**: p. 91–102.
177. Ferro, P., *et al.*, The influence of phase transformations on residual stresses induced by the welding process – 3D and 2D numerical models. *Modelling and Simulation in Materials Science and Engineering*, 2006. **14**: 117–136.
178. Runnemalm, H. and D. Berglund, Finite element simulation of welding prediction of deformation patterns. In *3rd TWI/EWI Workshop on Joining of Aerospace Materials*. 2000. Bokenäs, Sweden.
179. Dye, D., *et al.*, Modeling of the mechanical effects induced by the tungsten-inert-gas welding of the In718 superalloy. *Metallurgical and Materials Transactions A*, 2001. **32A**(July): 1713–1725.
180. Ravichandran, G., *et al.*, Prediction of axis shift distortion during circumferential welding of thin pipes using the finite element method. *Welding Journal*, 1997. **January**: 39s–55s.
181. Zhang, J., *et al.*, Modeling of weld residual stresses in core shroud structures. *Nuclear Engineering and Design*, 2000. **195**(2): 171–187.
182. Dong, Y., *et al.*, Finite element modeling of residual stresses in austenitic stainless steel pipe girth welds. *Welding Journal*, 1997. **76**(10): 442s–449s.
183. Fricke, S., E. Keim, and J. Schmidt. Fracture mechanics investigation of root formation and shrinkage during welding. In *ASME Pressure Vessel and Piping Conference*. 1998: San Diego, USA: ASME.
184. Fricke, S., E. Keim, and J. Schmidt. Modelling of root formation during the welding process with the help of the 3D FE method. In *Mathematical Modelling of Weld Phenomena 4*. 1998. Graz, Austria: The Institute of Materials.

185. Fricke, S., E. Keim, and J. Schmidt, Numerical weld modeling – a method for calculating weld-induced residual stresses. *Nuclear Engineering and Design*, 2001. **206**(2–3): 139–150.
186. Dike, J., *et al.*, Finite element modeling and validation of residual stresses in 304L girth welds. In *5th International Conference on Trends in Welding Research*. 1998. Pine Mountain, USA: ASM International.
187. Chakravarti, A., L. Mali, and J. Goldak. Prediction of distortion and residual stresses in panel welds. In *Computer Modelling of Fabrication Processes and Constitutive Behaviour of Metals*. 1986. Ottawa, Canada.
188. Tsai, C., S. Park, and W. Cheng, Welding distortion of a thin-plate panel structure. *Welding Journal*, 1999. **May**: 156s–165s.
189. Michaleris, P. and A. DeBiccari, Prediction of welding distortion. *Welding Journal*, 1997. **76**(4): 172s–181s.
190. Dong, P., J. Zhang, and M. Li, Computational modeling of weld residual stresses and distortions – an integrated framework and industrial applications. In *ASME Pressure Vessel and Piping Conference*. 1998. ASME Press. Vess. & Piping Division.
191. Rybicki, E., *et al.*, A finite element model for residual stresses in girth-butt welded pipes. In *Numerical modeling of manufacturing processes, ASME Winter Annual Meeting*, held in Atlanta, Georgia. 1977.
192. Rybicki, E.F., *et al.*, A finite-element model for residual stresses and deflections in girth-butt welded pipes. *ASME Journal of Pressure Vessel Technology*, 1978. **100**: 256–262.
193. Rybicki, E.F. and R.B. Stonesifer, Computation of residual stresses due to multipass welds in piping systems. *ASME Journal of Pressure Vessel Technology*, 1979. **101**(May): 149–154.
194. Ueda, Y. and K. Nakacho, Simplifying methods for analysis of transient and residual stresses and deformations due to multipass welding. *Transactions of Japan Welding Institute*, 1982. **11**(1): 95–103.
195. Leung, C., R. Pick, and D. Mok, Finite element analysis of multipass welds. *WRC Bulletin*, 1990. **356**(Aug): 11–33.
196. Hong, J., C. Tsai, and P. Dong, Assessment of numerical procedures for residual stress analysis of multipass welds. *Welding Journal*, 1998. **77**(9): 372s–382s.
197. Li, M., *et al.*, A 3D finite element analysis of temperature and stress fields in girth welded 304L stainless steel pipe. In *4th International Conference on Trends in Welding Research*. 1995. Gatlinburg, USA: ASM International.
198. Li, M., *et al.*, An experimental analysis of temperature and stress fields in girth welded 304L stainless steel pipes. In *4th International Conference on Trends in Welding Research*. 1995. Gatlinburg, USA: ASM International.
199. Ortega, A., *et al.*, Analysis and validation of multi-pass girth welds in pipes. In *8th International Conference of Casting, Welding and Advanced Solidification Processes*. 1998. San Diego, USA: The Minerals, Metals & Materials Society.
200. Dike, J., *et al.*, Finite element modeling of multipass GMA welds in steel plates. In *4th International Conference on Trends in Welding Research*. 1995. Gatlinburg, USA: ASM International.
201. Chakravarti, A., *et al.*, Prediction of distortion in overlayed repair welds. In *5th International Conference on Numerical Methods in Thermal Problems*. 1987. Montreal, Canada.
202. Hinrichsen, B., Prediction of welding induced distortion in thin-walled ship panels. *Ship Technology Research*, 1999. **46**: 153–163.

203. Huang, T., *et al.*, Fabrication and engineering technology for lightweight ship structures, Part 1: distortion and residual stresses in panel fabrication. *Journal of Ship Production*, 2004. **20**: 43–59.
204. Ueda, Y. and M. Yuan, Prediction of residual stresses in butt-welded plates using inherent strain. *ASME Journal of Engineering Materials and Technology*, 1993. **115**(Oct): 417–423.
205. Schimmöller, H., Bestimmung von Eigenspannungen in ebenen plattierten Werkstoffen, Teil 1: Rechnerische Grundlagen. *Materialprufung*, 1972. **14**(4): 115–122.
206. Schimmöller, H., Bestimmung von Eigenspannungen in ebenen plattierten Werkstoffen, Teil 2: Eigenspannungen in warmgewalzten und sprenggeschweissten austenit-plattierten Stahlblechen. *Materialprufung*, 1972. **14**(11): 380–387.
207. Yuan, M. and Y. Ueda, Prediction of residual stresses in welded T- and I-joints using inherent strains. *ASME Journal of Engineering Materials and Technology*, 1996. **118**: 229–234.
208. Jung, G. and C. Tsai, Plasticity-based distortion analysis for fillet welded thin-plate T-joints. *Welding Journal*, 2004. **June**: 177s–187s.
209. Lindgren, L.-E., Finite element modelling and simulation of welding, Part 3 Efficiency and integration. *Journal of Thermal Stresses*, 2001. **24**: 305–334.
210. Oberkampf, W.L. and T.G. Trucano, Verification and validation in computational fluid dynamics. *Progress in Aerospace Sciences*, 2002. **38**(3): 209–272.
211. Oberkampf, W.L., T.G. Trucano, and C. Hirsch, Verification, validation, and predictive capability in computational engineering and physics. *Applied Mechanics Reviews*, 2004. **57**(5): 345–384.
212. Oberkampf, W.L., *et al.*, Error and uncertainty in modeling and simulation. *Reliability Engineering and System Safety*, 2002. **75**(3): 333–357.
213. Oberkampf, W.L., *et al.*, Challenge problems: uncertainty in system response given uncertain parameters. *Reliability Engineering and System Safety*, 2004. **85**(1–3): 11–19.
214. Meng, X.N. and T.A. Laursen, Energy consistent algorithms for dynamic finite deformation plasticity. *Computer Methods in Applied Mechanics and Engineering*, 2002. **191**(15–16): 1639–1675.
215. Sudnik, W., D. Radaj, and W. Erofeew, Validation of computerised simulation of welding processes. In *Mathematical Modelling of Weld Phenomena 4*. 1998. Seggau, Graz: The Institute of Materials, Minerals and Mining.
216. Alberg, H., *Simulation of Welding and Heat Treatment. Modelling and Validation.* 2005: Department of Applied Physics and Mechanical Engineering, Luleå University of Technology.
217. Belytschko, T. and K. Mish, Computability in non-linear solid mechanics. *International Journal of Numerical Methods of Engineering*, 2001. **52**: 3–21.
218. Trucano, T.G., *et al.*, Calibration, validation, and sensitivity analysis: What's what. *Reliability Engineering and System Safety*, 2006. **91**(10–11): 1331–1357.
219. Radaj, D. and L.-E. Lindgren, Verification and validation in computational welding mechanics. In *Mathematical Modelling of Weld Phenomena 8*. 2006. Seggau, Graz: The Institute of Materials, Minerals and Mining.
220. Kajberg, J., *et al.*, High strain-rate tensile testing and viscoplastic parameter identification using microscopic high-speed photography. *International Journal of Plasticity*, 2004. **20**(4–5): 561–575.
221. Garcia de Andrés, C., *et al.*, Modelling of kinetics and dilatometric behaviour of non-isothermal pearlite-to-austenite transformation in an euctectoid steel. *Scripta Materialia*, 1998. **39**(6): 791–796.

222. Cavallo, N., *et al.*, Thermomechanical behaviour of a carbon manganese steel under martensitic transformation. In *The Fifth International Conference on Residual Stresses*. 1997. Linköping, Sweden: Linköping University.
223. Taleb, L., N. Cavallo, and F. Waeckel, Experimental analysis of transformation plasticity. *International Journal of Plasticity*, 2001. **17**(1): 1–20.
224. Taleb, L., N. Cavallo, and F. Waeckel, Erratum to 'Experimental analysis of transformation plasticity' [*International Journal of Plasticity*, 17 (2001) 1–20]. *International Journal of Plasticity*, 2001. **17**(7): 1029.
225. Otsuka, T., Y. Wakasu, and T. Inoue, A simple identification of transformation plastic behaviour and some data for heat treating materials. *International Journal of Materials and Product Technology*, 2005. **24**(1–4): 298–311.
226. Alberg, H. and D. Berglund, Comparison of plastic, viscoplastic, and creep models when modelling welding and stress relief heat treatment. *Computer Methods in Applied Mechanics and Engineering*, 2003. **192**(49–50): 5189–5208.
227. Lundbäck, A., H. Alberg, and P. Henriksson, Simulation and validation of TIG-welding and post weld heat treatment of an Inconel 718 plate. In *Mathematical Modelling of Weld Phenomena 7*. 2003. Graz, Austria: The Institute of Materials.
228. Sudnik, W., D. Radaj, and W. Erofeew, Computer simulation of laser beam welding, modelling and verification. *Journal of Physics D: Applied Physics*, 1996. **29**: 2811–2817.
229. Ortega, A., *et al.*, Thermomechanical modeling of a stationary gas metal arc weld: a comparison between numerical and experimental results. In *3rd International Conference on Trends in Welding Research*. 1992. Gatlinburg, USA: ASM International.
230. Wikman, B., *Inverse Modelling for Determination of Constitutive Parameters*. 1999: Department of Mechanical Engineering: Luleå, p. 47.
231. Bruhns, O.T. and D.K. Anding, On the simultaneous estimation of model parameters used in constitutive laws for inelastic material behaviour. *International Journal of Plasticity*, 1999. **15**(12): 1311–1340.
232. Lindgren, L.-E., H. Alberg, and K. Domkin, Constitutive modelling and parameter optimisation. In *VII International Conference on Computational Plasticity COMPLAS 2003*. 2003. Barcelona.
233. Mahnken, R. and E. Stein, Parameter identification for viscoplastic models based on analytical derivatives of a least-squares functional and stability investigation. *International Journal of Plasticity*, 1996. **12**(4): 451–479.
234. Zentar, R., P. Hicher, and G. Moulin, Identification of soil parameters by inverse analysis. *Computer and Geotechnics*, 2001. **28**: 129–144.
235. Saleb, A., *et al.*, A general hereditary multimechanism-based deformation model with application to the viscoelastoplastic response of titanium alloys. *International Journal of Plasticity*, 2001. **17**: 1305–1350.
236. Mahnken, R. and E. Stein, The identification of parameters for visco-plastic models via finite-element methods and gradient methods. *Modelling and Simulations in Materials Science Engineering*, 1994. **2**: 597–616.
237. Eriksson, M., B. Wikman, and G. Bergman, Estimation of material parameters at elevated temperatures by inverse modelling of a gleeble experiment. *Modelling and Simulations in Materials Science Engineering*, 2002. **10**: 277–284.
238. Mahnken, R. and E. Stein, A unified approach for parameter identification of inelastic material models in the frame of the finite element method. *Computer Methods in Applied Mechanics and Engineering*, 1996. **136**: 225–258.
239. Mahnken, R. and E. Kuhl, Parameter identification of gradient enhanced damage

models with the finite element method. *European Journal of Mechanics A Solids*, 1999. **18**: 819–835.

240. Fletcher, R., *Practical Methods of Optimization, Vol. 1 Unconstrained Optimization*. 1980: John Wiley & Sons.
241. Fletcher, R., *Practical Methods of Optimization, Vol. 2 Constrained Optimization*. 1981: John Wiley & Sons.
242. Gen, M. and R. Cheng, *Genetic Algorithms & Engineering Optimzation*. Wiley Series in Engineering Design and Automation, ed. H. Parsaei. 2000: John Wiley & Sons.
243. Muller, D. and G. Hartmann, Identification of materials parameters for inelastic constitutive models using principles of biologic evolution. *ASME Journal of Engineering Materials and Technology*, 1989. **111**(3): 299–305.
244. Furukawa, T., *et al.*, An automated system for simulation and parameter identification of inelastic constitutive models. *Computer Methods in Applied Mechanics and Engineering*, 2002. **191**: 2235–2260.
245. Huber, N. and C. Tsakmakis, A neural network tool for identifying the material parameters of a finite deformation viscoplasticity model with static recovery. *Computer Methods in Applied Mechanics and Engineering*, 2001. **191**: 353–384.
246. Berglund, D., Validation of Models for Welding and Post Weld Heat Treatment in Product Development of Aerospace Components. 2003: Department of Applied Physics and Mechanical Engineering, Luleå University of Technology.
247. Stouffer, D.C. and L.T. Dame, *Inelastic Deformations of Metals – Models, Mechanical Properties and Metallurgy*. 1996: John Wiley & Sons.
248. Fischer, F.D., *et al.*, Transformation induced plasticity revised and updated formulations. *International Journal of Solids and Structures*, 1998. **35**(18): 2209–2227.
249. Bell, J., *The Experimental Foundation of Solid Mechanics*. Mechanics of Solids Series, ed. C. Truesdell, Vol. I. 1984: Springer-Verlag.
250. Bodner, S., *Unified Plasticity for Engineering Applications*. Mathematical Concepts and Methods in Science and Engineering Series, ed. A. Miele. Vol. 47. 2002: Kluwer Academic/Plenum Publishers.
251. ASM International Handbook Committee, ed., *Properties and Selection: Irons, Steels and High-performance Alloys*, 10th edn, *Metals Handbook*, Vol. 1. 1990, ASM International.
252. ASM International Handbook Committee, ed., *Stainless Steels*, ASM Specialty Handbook, ed. J. Davis. 1994, ASM International.
253. Jäniche, W., *et al.*, eds, *Steel A Handbook for Materials Research and Engineering. Volume 2: Applications*. 1992: Verlag Stahleisen.
254. Richter, F., *Die wichtigsten physikalischen Eigenschaften von 52 Eisenwerkstoffen*. 1973: Verlag Stahleisen.
255. Hepworth, J., Finite element calculation of residual stresses in welds. In *International Conference for Non-linear Problems*. 1980. Swansea, UK.
256. Ueda, Y., *et al.*, Mathematical treatment of phase transformation and analytical calculation of restraint stress-strain. *Transactions of Japan Welding Institute*, 1985. **14**(1): 153–162.
257. Free, J. and R. Porter Goff, Predicting residual stresses in multi-pass weldments with the finite element method. *Computers and Structures*, 1989. **32**(2): 365–378.
258. Tekriwal, P. and J. Mazumder, Transient and residual thermal strain-stress analysis of GMAW. *ASME Journal of Engineering Materials and Technology*, 1991. **113**: 336–343.

259. Ueda, Y., K. Nakacho, and T. Shimuzu, Improvement of residual stresses of circumferential joint of pipe by heat-sink welding. *ASME Journal of Pressure Vessel Technology*, 1986. **108**(2): 14–23.
260. Brown, S. and H. Song, Finite element simulation of welding of large structures. *Journal of Engineering for Industry*, 1992. **114**: 441–451.
261. Brown, S. and H. Song, Implications of three-dimensional numerical simulations of welding of large structures. *Welding Journal*, 1992. **February**: 55s–62s.
262. Nickell, R. and H. Hibbitt, Thermal and mechanical analysis of welded structures. *Nuclear Engineering and Design*, 1975. **32**: 110–120.
263. Ueda, Y., *et al.*, Finite element modeling and residual stress calculation for multi-pass single welded joint between a plate and the penetrating pipe. In *5th International Conference on Modeling of Casting, Welding and Advanced Solidification Processes*. 1991. Davos, Austria: The Minerals, Metals & Materials Society.
264. Shim, Y., *et al.*, Determination of residual stress in thick-section weldments. *Welding Journal*, 1992. **71**: 305s–312s.
265. Michaleris, P., *et al.*, Incorporation of residual stress effects into fracture mechanics assessments via the finite element method. In *Fatigue and Fracture Mechanics*. 1996. Saratoga Springs, USA: ASTM.
266. Ueda, Y., *et al.*, Transient and residual stresses from multipass welding in very thick plates and their reduction from stress relief annealing. *Transactions of Japan Welding Research Institute*, 1976. **5**: 179–187.
267. Ueda, Y., *et al.*, Transient and residual stresses from multipass welding in very thick plates and their reduction from stress relief annealing. In *3rd International Conference on Pressure Vessel Technology*. 1977. Tokyo.
268. Josefson, B., Residual stresses and their redistribution during annealing of a girth-butt welded thin-walled pipe. *ASME Journal of Pressure Vessel Technology*, 1982. **104**: 245–250.
269. Inoue, T. and Z. Wang, High temperature behaviour of steels with phase transformation and the simulation of quenching and welding processes. In *4th International Conference on the Mechanical Behaviour of Materials*. 1983. Stockholm, Sweden.
270. Wang, Z. and T. Inoue, A viscoplastic constitutive relationship with phase transformation and the application to the process of welding. In *4th International Conference on Mechanical Behaviour of Materials*. 1983. Stockholm, Sweden.
271. Wang, Z. and T. Inoue, Viscoplastic constitutive relation incorporating phase transformation – application to welding. *Material Science and Technology*, 1985. **1**(Oct): 899–903.
272. Bergheau, J. and J. Leblond, Coupling between heat flow, metallurgy and stress–strain computations in steels: the approach developed in the computer code Sysweld for welding or quenching. In *5th International Conference on Modeling of Casting, Welding and Advanced Solidification Processes*. 1991. Davos, Austria: The Minerals, Metals & Materials Society.
273. Devaux, J., D. Pont, and J. Leblond. Numerical simulation of the repair of a defect-containing zone by a manual welding procedure. In *ASME 10th Conference on Offshore and Arctic Engineering (OMAE)*. 1991. Stavanger, Norway.
274. Inoue, T., Metallo-thermo-mechanics. *Transactions of Japan Welding Research Institute*, 1996. **25**(2): 69–87.
275. Inoue, T., ed. Residual stresses and distortion metallo-thermo-mechanics simulation of engineering processes incorporating phase transformation. In *Mathematical Modelling of Weld Phenomena 4*, H. Cerjak and H. Bhadeshia, Editors. 1998,

Institute of Materials, pp. 547–575.

276. Roelens, J.-B., F. Maltrud, and J. Lu, Determination of residual stresses in submerged arc multi-pass welds by means of numerical simulation and comparison with experimental measurements. *Welding in the World*, 1994. **33**(3): 152–159.
277. Roelens, J.-B., Numerical simulation of multipass submerged arc welding – determination of residual stresses and comparison with experimental measurements. *Welding in the World*, 1995. **35**(2): 110–117.
278. Börjesson, L., Coupled thermal, metallurgical and mechanical models of multipass welding. In *Mechanical Engineering*. 1999: Luleå University of Technology.
279. Myhr, O.R. and O. Grøng, Modelling of non-isothermal transformations in alloys containing a particle distribution. *Acta Materialia*, 2000. **48**(7): 1605–1615.
280. Myhr, O.R., O. Grøng, and S.J. Andersen, Modelling of the age hardening behaviour of Al-Mg-Si alloys. *Acta Materialia*, 2001. **49**(1): 65–75.
281. Dufrene, L., J. Truc, and R. Martin, Modeling of thermal stresses during an electron-beam welding process. *Journal of Thermal Stresses*, 1996. **19**(3): 211–238.
282. Sjöström, S., *The Calculation of Quench Stresses in Steel*. 1982: University of Linköping.
283. Perlade, A., O. Bouaziz, and Q. Furnemont, A physically based model for TRIP-aided carbon steels behaviour. *Materials Science and Engineering A*, 2003. **356**: 145–152.
284. Gladman, T., I.D. McIvor, and F. Pickering, Some aspects of the structure-property relationships in high-carbon ferrite-pearlite steels. *Journal of Iron and Steel Institute*, 1972. **210**: 916–930.
285. Bouquerel, J., K. Verbeken, and B.C. De Cooman, Microstructure-based model for the static mechanical behaviour of multiphase steels. *Acta Materialia*, 2006. **54**(6): 1443–1456.
286. Leblond, J., G. Mottet, and J. Devaux, A theoretical and numerical approach to the plastic behavior of steels during phase transformations – II. Study of classical plasticity for ideal-plastic phases. *Journal of the Mechanics and Physics of Solids*, 1986. **34**(4): 411–432.
287. Bouaziz, O. and P. Buessler, Iso-work increment assumption for heterogeneous material behaviour modelling. *Advanced Engineering Materials*, 2004. **6**(1–2): 79–83.
288. Brickstad, B. and B. Josefson, A parametric study of residual stresses in multi-pass butt-welded stainless steel pipes. *International Journal of Pressure Vessels and Piping*, 1998. **75**: 11–25.
289. Mahin, K., *et al.*, Prediction and measurements o residual elastic strain distributions in gas tungsten arc welds. *Welding Journal*, 1991. **70**(9): 245s–260s.
290. Mahin, K., *et al.*, Evaluation of residual stress distribution in a traveling GTA weld using finite element and experimental techniques. In *Modeling of Casting and Welding Processes IV*. 1988. Sheraton Palm Coast, Florida, USA: The Minerals, Metals & Materials Society.
291. Brust, F., P. Dong, and J. Zhang, A constitutive model for welding process simulation using finite element methods. In *Advances in Computational Engineering Science*. 1997: Tech Science Press, Anaheim, CA.
292. Lindgren, L.-E., Finite element modelling and simulation of welding, Part 2 Improved material modelling. *Journal of Thermal Stresses*, 2001. **24**: 195–231.
293. Moore, J., M. Bibby, and J. Goldak, A comparison of the point source and finite element schemes for computing weld cooling. In *Welding Research: The State of the Art*. 1985. Toronto, Canada: ASM.

294. Pehlke, R., A. Jeyarajan, and H. Wada, *Summary of Thermal Properties for Casting Alloys and Mold Materials*. 1982: University of Michigan.
295. Dubois, D., J. Devaux, and J.B. Leblond, Numerical simulation of a welding operation: Calculation of residual stresses and hydrogen diffusion. In *Fifth International Conference on Pressure Vessel Technology*. 1984. San Francisco.
296. Ashby, M., Physical modelling of material problems. *Materials Science and Technology*, 1992. **8**(Feb): 102–111.
297. Bru, D., *et al.*, Influence of material properties at high temperatures on the modelling of welding residual stress and deformation state. In *Mathematical Modelling of Weld Phenomena 3*. 1997. Graz, Austria: The Institute of Materials.
298. Sekhar, N., *et al.*, Characterisation of distortion during the laser welding of wasp-alloy: modelling and in-situ process monitoring. In *5th International Conference on Trends in Welding Research*. 1998. Pine Mountain, USA: ASM International.
299. Suzuki, H., *et al.*, *Studies on the Flow Stress of Metals and Alloys*. 1968: The Institute of Industrial Science, University of Tokyo.
300. Argyris, J.H., J. Szimmat, and K.J. Willam, Computational aspects of welding stress analysis. *Computer Methods in Applied Mechanics and Engineering*, 1982. **33**: 653–666.
301. Argyris, J., J. Szimmat, and K. Willam, Finite element analysis of arc-welding processes. In *Numerical Methods in Heat Transfer*, R. Lewis, Editor. 1985: Wiley, pp. 35–58.
302. Chidiac, S. and F. Mirza, Thermal stress analysis due to welding processes by the finite element method. *Computers and Structures*, 1993. **46**(3): 407–412.
303. Myhr, O., *et al.*, Modeling of microstructure evolution, residual stresses in processing and distortions in 6082-T6 aluminium weldments. *Welding Journal*, 1998. **77**(7): 286s–292s.
304. Myhr, O., *et al.*, Modeling of microstructure evolution and residual stresses in processing and welding of 6082 and 7108 aluminium alloys. In *5th International Conference on Trends in Welding Research*. 1998. Pine Mountain, Georgia, USA.
305. Goldak, J., *et al.*, Thermal stress analysis of welds: From melting point to room temperature. *Transactions of Japan Welding Research Institute*, 1996. **25**(2): 185–189.
306. Goldak, J., *et al.*, Thermal stress analysis in solids near the liquid region in welds. In *Mathematical Modelling of Weld Phenomena 3*. 1997. Graz, Austria: The Institute of Materials.
307. Winters, W. and K. Mahin, PASTA2D Modeling of an axisymmetric stationary gas tungsten arc weld. in *5th International Conference on Modeling of Casting, Welding and Advanced Solidification Processes*. 1991. Davos, Austria: The Minerals, Metals & Materials Society.
308. Bammann, D.J., Modeling temperature and strain rate dependent large deformations of metals. *Applied Mechanics Review*, 1990. **43**(5): 312–319.
309. Bammann, D.J., M.L. Chiesa, and G.C. Johnson, Modeling large deformation and failure in manufacturing processes. In *19th International Congress on Theoretical and Applied Mechanics*. 1996. Kyoto: Elsevier Science.
310. Chen, Y. and I. Sheng, Residual stress in weldment. *Journal of Thermal Stresses*, 1992. **15**(1): 53–69.
311. Bammann, D.J. and A.R. Ortega, The influence of the Bauschinger effect and yield definition on the modeling of welding processes. In *Modeling of Casting, Welding and Advanced Solidification Processes VI*. 1993. Warrendale: The Minerals, Metals & Materials Society.

312. Murthy, Y., G. Venkata Rao, and P. Krishna Iyer, Analysis of residual stresses in hemispherical head to cylindrical shell joints of steam generator by weld cycle simulation. *Journal of Materials Processing Technology*, 1994. **44**(3–4): 273–280.
313. Fischer, F.D., G. Resiner, E. Werner, K. Tanaka, G. Cailletaud, and T. Antretter, A new view on transformation induced plasticity (TRIP). *International Journal of Plasticity*, 2000. **16**(7–8): 723–748.
314. Greenwood, G. and R. Johnson, The deformation of metals under small stresses during phase transformation. *Proceddings of the Royal Society*, 1965. **283**: 403–422.
315. Magee, C. and R. Davies, On the volume expansion accompanying the fcc to bcc transformation in ferrous alloys. *ACTA Metallurgica*, 1972. **20**: 1031–1043.
316. Oddy, A., J. Goldak, and M. McDill, Transformation effects in the 3D finite element analysis of welds. In *The 2nd International Conference on Trends in Welding Research*. 1989. Gatlinburg, USA.
317. Sjöström, S., Interactions and constitutive models for the quench stresses in steel. In *International Symposium on Calculation of Internal Stresses in Heat Treatment of Metallic Materials*. 1984. Linköping, Sweden.
318. Sjöström, S., Interactions and constitutive models for calculating quench stresses in steel. *Materials Science and Technology*, 1985. **1**(Oct): 823–829.
319. Vincent, Y., J.-M. Bergheau, and J.-B. Leblond, Viscoplastic behaviour of steels during phase transformations. *Comptes Rendus Mecanique*, 2003. **331**: 587–594.
320. Taleb, L. and S. Petit, New investigations on transformation induced plasticity and its interaction with classical plasticity. *International Journal of Plasticity*, 2006. **22**(1): 10–130.
321. Oddy, A., J. Goldak, and R. Reed, Martensite formation, transformation plasticity and stress in high strength steel welds. In *3rd International Conference on Trends in Welding Research*. 1992. Gatlinburg, USA: ASM International.
322. Antretter, T., *et al.*, Theory, experiments and numerical modelling of phase transformations with emphasis on TRIP. In *International Conference on TRIP-aided High Strength Ferrous Alloys*. 2006. Ghent: Wissenschaftsverlag Mainz GmbH.
323. Koistinen, D. and R. Marburger, A general equation prescribing the extent of the austenite-martensite transformation in pure-carbon alloys amd plain carbon steels. *Acta Metallica*, 1959. **7**: 59–60.
324. Leblond, J., G. Mottet, and J. Devaux, A theoretical and numerical approach to the plastic behavior of steels during phase transformations – I. Derivation of general relations. *Journal of the Mechanics and Physics of Solids*, 1985. **34**(4): 395–409.
325. Leblond, J.B., J. Devaux, and J.C. Devaux, Mathematical modelling of transformation plasticity in steels I: Case of ideal-plastic phases. *International Journal of Plasticity*, 1989. **5**: 551–572.
326. Leblond, J.B., Mathematical modelling of transformation plasticity in steels II: Coupling with strain hardening phenomena. *International Journal of Plasticity*, 1989. **5**: 573–591.
327. Leblond, J., *et al.*, Metallurgical and mechanical consequences of phase transformations in numerical simulations of welding processes. In *Modeling in Welding, Hot Powder Forming, and Casting*, L. Karlsson, Editor. 1997: ASM International, pp. 61–89.
328. Devaux, J., *Comportement plastique des aciers en cours de transformations de phases. Etude numérique des lois de mélange et de la plasticité de transformation*. 1998: SYSTUS International.

329. Leblond, J., *et al.*, Plastic behaviour of steels during phase transformations. In *IUTAM Symposium on Thermomechanical Couplings in Solids*. 1987: Elsevier Science Publisher BV.
330. Leblond, J. and J. Devaux, Interactions between transformation plasticity and strain hardening phenomena in steels. In *2nd International Conference on Residual Stresses ICRS2*. 1988. Nancy, France.
331. Mitter, W., Umwandlungsplastizität und ihre Berucksichtung bei der Berechnung von Eigenspannungen. In *Materialkundlich-Technische Reihe 7*. 1987, Bornträger-Verlag.
332. Berveiller, M. and F. Fischer, eds, *Mechanics of Solids with Phase Changes*. CISM International Centre for Mechanical Sciences. Vol. 368. 1997, Springer.
333. Thiessen, R.G., I.M. Richardson, and J. Sietsma, Physically based modelling of phase transformations during welding of low-carbon steel. *Materials Science and Engineering: A*, 2006. **427**(1–2): 223–231.
334. International, A., *One-Minute-Mentor*. 2006. http://www.asminternational.org/Content/NavigationMenu/Magazines/HeatTransferProgress/One_Minute_Mentor1/One_Minute_Mentor.htm
335. Leblond, J.B. and J. Devaux, A new kinetic model for anisothermal metallurgical transformations in steels including effect of austenite grain size. *Acta Metallurgica*, 1984. **32**(1): 137–146.
336. Leblond, J.-B., *et al.*, Mathematical models of anisothermal phase transformations in steels, and predicted plastic behaviour. *Materials Science and Technology*, 1985. **1**(October): 815–822.
337. Johnson, W. and R. Mehl, Reaction kinetics in processes of nucleation and growth. *Transactions of American Institute of Mining and Metallurgical Engineers*, 1939. **135**: 416–458.
338. Avrami, M., Kinetics of phase change. III: Granulation, phase change and microstructures. *Journal of Chemical Physics*, 1941. **9**: 177–184.
339. Kolmogorov, A., Statistical theory of crystallization of metals (in Russian). *Izvestia Akademia Nauk SSSR Ser. Mathematica*, 1937. **1**(1): 355–359.
340. Porter, D. and K. Easterling, *Phase Transformations in Metals and Alloys*, 2nd edn. 1992: Chapman & Hall.
341. Kang, S.-H. and Y.-T. Im, Three-dimensional finite-element analysis of the quenching process of plain-carbon steel with phase transformation. *Metallurgical and Materials Transactions*, 2005. **36A**(9): 2315.
342. Fernandes, F., S. Denis, and A. Simon, Mathematical model coupling phase transformation and temperature evolution during quenching of steels. *Materials Science and Technology*, 1985. **1**(Oct): 838–844.
343. Hildenwall, B., Prediction of the residual stresses created during quenching. In *Linköping Studies in Science and Technology*. 1979. Linköping University.
344. Kirkaldy, J.S. and D. Venugopalan, Prediction of microstructure and hardenability in low alloy steels. In *International Conference on Phase Transformations in Ferrous Alloys*. 1983. Philadelphia.
345. Kirkaldy, J.S. and D. Venugopalan, Phase transformations in ferrous alloys. In *International Conference on Phase Transformations in Ferrous Alloys*. TMS-AIME, Warrendale, 1984, pp. 125–148.
346. Kirkaldy, J.S. and R.C. Sharma, A new phenomenology for steel IT and CCT curves. *Scripta Metallurgica*, 1982. **16**: 1193–1198.
347. Watt, D., *et al.*, An algorithm for modelling microstructural development in weld heat-affected zones (Part A) Reaction kinetics. *Acta Metallurgica*, 1988. **36**(11):

3029–3035.
348. Henwood, C., *et al.*, Coupled transient heat transfer – microstructure weld computations (Part B). *Acta Metallurgica*, 1988. **36**(1): 3037–3046.
349. Oddy, A.S., J.M.J. McDill, and L. Karlsson, Microstructural predictions including arbitrary thermal histories, reaustenization and carbon segregation effects. *Canadian Metallurgical Quarterly*, 1996. **35**(3): 275–283.
350. Kirkaldy, J., B. Thomson, and E. Baganis, Prediction of multicomponent equilibrium and transformation diagrams for low alloy steels. In *Proceedings of Symposium on Hardenability Concepts with Applications to Steel*, Chicago, USA. 1977: The Metallurgical Society AIME, Heat Treatment Committee/American Society for Metals Activity on Phase Transformations.
351. Pickering, F., Physical metallurgical development of stainless steels. In *Stainless Steels 84*. 1984. Gothenburg, Sweden.
352. Easterling, K., M. Ashby, and A. Li, A first report for grain growth in welds. *Acta Metallurgica*, 1982. **30**: 1969–1978.
353. Svensson, L.-E., *Control of Microstructures and Properties in Steel Arc Welds*. Materials Science and Technology Series, ed. B. Ralph. 1994: CRC Press.
354. Evans, G. and N. Bailey, *Metallurgy of Basic Weld Metal*. 1997: Woodhead Publishing.
355. Olson, G. and M. Cohen, Stress-assisted isothermal martensitic transformation; application to TRIP steels. *Metallurgical Transactions*, 1982. **13A**(Nov): 1907–1914.
356. Tsakiris, V. and D.V. Edmonds, Martensite and deformation twinning in austenitic steels. *Materials Science and Engineering A*, 1999. **273–275**: 430–436.
357. Ganesh Sundara Raman, S. and K. Padmanabhan, Tensile deformation-induced martensitic transformation in AISI 304LN austenitici stainless steel. *Journal of Materials Science Letters*, 1994. **13**(5): 389–392.
358. Reed-Hill, R.E. and R. Abbaschian, *Physical Metallurgy Principles*, 3rd edn. 1992: PWS-KENT Publishing Company.
359. Fischer, F., A micromechanical model for transformation plasticity in steels. *Acta Metallurgical et Materialia*, 1990. **38**(8): 1535–1546.
360. Tanaka, K. and Y. Sato, A mechanical view of transformation plasticity. *Ingeniur-Archiv*, 1985. **55**: 147–155.
361. Inoue, T. and Z. Wang, Finite element analysis of coupled thermo-inelastic problem with phase transformation. In J. Wood, H. David, and J. Alexander (Editors), *Numerical Methods in Industrial Forming Processes*. 1982. Pineridge Press, Swansea, pp. 391–400.
362. Inoue, T. and Z. Wang, Coupling between stress, temperature, and metallic structures during processes involving phase transformations. *Materials Science Technology*, 1985. **1**: 845–850.
363. Denis, S., *et al.*, Stress-phase-transformation interactions – basic principles, and calculation of internal stresses. *Materials Science and Technology*, 1985. **1**(Oct): 805–814.
364. Kharkin, V., *et al.*, Effects of latent heat of fusion on thermal processes during welding. In *Mathematical Modelling of Weld Phenomena 7*. 2003. Seggau, Austria: Teschnische Universität Graz.
365. Feng, Z., A computational analysis of thermal and mechanical conditions for weld solidification cracking. *Welding in the World*, 1994. **33**(5): 340–347.
366. Feng, Z., T. Zacharia, and S. David, On the thermomechanical conditions for weld metal solidification cracking. In *Mathematical Modelling of Weld Phenomena 3*.

1997. Graz, Austria: The Institute of Materials.

367. Dike, J., J. Brooks, and J. Krafcik, Finite element modeling and verification of thermal-mechanical behavior in the weld pool region. In *4th International Conference on Trends in Welding Research*. 1995. Gatlinburg, USA: ASM International.
368. Madhusudana, C. and L. Fletcher, Contact heat transfer – the last decade. *AIAA Journal*, 1985. **24**(3): 510–523.
369. Salgon, J., *et al.*, A mechanical and geometrical approach to thermal contact resistance. *International Journal of Heat Mass Transfer*, 1997. **40**(5): 1121–1129.
370. Malinowski, Z., J. Lenard, and M. Davies, A study of the heat-transfer coefficient as a function of temperature and pressure. *Journal of Materials Processing*, 1994. **41**: 125–142.
371. Lambert, M. and L. Fletcher, Review of models for thermal contact conductance of metals. *Journal of Thermophysics and Heat Transfer*, 1997. **11**(2): 129–140.
372. Incropera, F. and D. DeWitt, *Fundamentals of Heat and Mass Transfer*, 3rd edn. 1990: John Wiley.
373. Bamberger, M. and B. Prinz, Determination of heat transfer coefficients during water cooling of metals. *Materials Science and Technology*, 1986. **2**(4): 410–415.
374. Goldak, J., Modeling thermal stresses and distortions in welds. In *2nd International Conference on Trends in Welding Research*. 1989. Gatlinburg, USA: ASM International.
375. Goldak, J., *et al.*, Heat and fluid flow in welds. In *International Welding Congress on Joining Research*, 1990. Montreal, Canada.
376. Goldak, J., Thermal analysis of welds. In *Modeling in Welding, Hot Powder Forming and Casting*, K. Karlsson, Editor. 1997. ASM International, Ohio, pp. 17–29.
377. Goldak, J., *et al.*, Computer modeling of heat flow in welds. *Metallurgical Transactions B*, 1986. **17B**: 587–600.
378. Goldak, J., *et al.*, Computational heat transfer for weld mechanics. In *Trends in Welding Research*. 1986. Gatlinburg, pp. 15–20.
379. Lindgren, L.E., Numerical modelling of welding. *Computer Methods in Applied Mechanics and Engineering*, 2006. **195**(48–49): 6710–6736.
380. Paley, Z. and P. Hibbert, Computation of temperatures in actual weld design. *Welding Journal*, 1975. **54**: 385s–392s.
381. Friedman, E., Analysis of weld puddle distortion and its effect on penetration. *Welding Journal*, 1978. **June**: 161s–166s.
382. Goldak, J., A. Chakravarti, and M. Bibby, A new finite element model for welding heat sources. *Metallurgical Transactions B*, 1984. **15B**: 299–305.
383. Sabapathy, P.N., M.A. Wahab, and M.J. Painter, The prediction of burn-through during in-service welding of gas pipelines. *International Journal of Pressure Vessels and Piping*, 2000. **77**(11): 669–677.
384. Das, S., *et al.*, Finite element modeling of a single-pass GMA weldment. In *Modeling of Casting, Welding and Advanced Solidification Processes VI*. 1993: The Minerals, Metals & Materials Society.
385. Lundbäck, A., CAD-support for heat input in FE-model. In *Mathematical Modelling of Weld Phenomena 6*. 2002. Graz, Austria: The Institute of Materials.
386. Carmet, A., *et al.*, Experimental and numerical study of residual stresses and strains in an electron-beam-welded joint. In *International Conference on Residual Stresses (ICRS2)*. 1988. Nancy.
387. Jones, B., A. Emery, and J. Marburger, An analytical and experimental study of the

effects of welding parameters in fusion welds. *Welding Journal*, 1993. **72**(2): 51s–59s.

388. Jones, B., A. Emery, and J. Marburger, Design and analysis of test coupon for fusion welding. *ASME Journal of Pressure Vessel and Technology*, 1993. **115**: 38–46.
389. Lindgren, L.-E., Welding simulation – 'knowing what isn't necessary to know'. In *Mathematical Modelling of Weld Phenomena 6*. 2001. Graz, Austria: The Institute of Materials, Minerals and Mining.
390. Berglund, D. and H. Runnemalm, Comparison of deformation pattern and residual stresses in finite element models of a TIG-welded stainless steel plate. In *6th International Conference on Trends in Welding Research*. 2002. Pine Mountain, Georgia: ASM International.
391. Kussmaul, K., *et al.*, A contribution to the numerical and experimental determination of residual stresses in welds. *Nuclear Engineering*, 1989. **112**: 337–348.
392. Lindgren, L.-E., A welding simulation revisited. In J.J. Skrzypek and R.B. Hetnarski (Editors), *Third International Congress on Thermal Stresses*. 1999. Krakow University, Krakow, Poland.
393. Parkus, H., *Thermoelasticity*, 2nd edn. 1968: Springer-Verlag.
394. Barrentine, L., *Design of Experiments. A Simplified Approach.* 1999: ASQ Quality Press.
395. Gep, B., W. Hunter, and J. Hunter, *Statistics for Experimenters: An Introduction to Design, Data Analysis and Model Building*. 1978: Wiley-Interscience.
396. Olabi, A.G., *et al.*, Minimisation of the residual stress in the heat affected zone by means of numerical methods. *Materials and Design*, in press, available online 24 October 2006. http://www.sciencedirect.com/science?_ob=ArticleURL&_udi=B6TX5-4M644Y1-1&_user=650452&_coverDate=10%2F24%2F2006&_alid=562896255&_rdoc=3&_fmt=full&_orig=search&_cdi=5581&_sort=d&_docanchor=&view=c&_ct=3&_acct=C000035038&_version=1&_urlVersion=0&_userid=650452&md5=8595df15d467546cf51f2714ec8d9dd2
397. Casalino, G., F. Curcio, and F. Memola Capece Minutolo, Investigation on Ti6Al4V laser welding using statistical and Taguchi approaches. *Journal of Materials Processing Technology*, 2005. **167**(2–3): 422–428.
398. Wriggers, P., W. Wagner, and C. Miehe, A quadratically convergent procedure for the calculation of stability points in finite element analysis. *Computer Methods in Applied Mechanics and Engineering*, 1988. **70**(3): 329–347.
399. Jonsson, M., L. Karlsson, and L.-E. Lindgren, Deformations and stresses in butt-welding of large plates with special reference to the mechanical material properties. *Journal of Engineering Materials and Technology*, 1985. **107**(10): 265–270.
400. Saint-Germain, B., P. Pilvin, and F. Gabriel, Comparison of heat source models and constitutive laws in finite element modelling of welding distortions confrontation to a TIG fusion line of an austenitit steel thin plate. In *Mathematical Modelling of Weld Phenomena 7*. 2003. Graz, Austria.
401. Zhu, X.K. and Y.J. Chao, Effects of temperature-dependent material properties on welding simulation. *Computers and Structures*, 2002. Only available online. http://www.sciencedirect.com/science?_ob=ArticleURL&_udi=B6TX5-4M644Y1-1&_user=650452&_coverDate=10%2F24%2F2006&_alid=562082244&_rdoc=3&_fmt=summary&_orig=search&_cdi=5581&_sort=d&_docanchor=&view=c&_ct=3&_acct=C000035038&_version=1&_urlVersion=0&_userid=650452&md5=d0e15a9a6300f1b097af8ca8b33baeb8
402. Michaleris, P., Prediction of welding distortion. *Welding Journal*, 1997. **76**: 172s.

403. Deo, M. and P. Michaleris, Elimination of welding induced buckling distortion. In *6th International Conference on Trends in Welding Research*. 2002. Pine Mountain, Georgia: ASM International.
404. Deo, M., J. Sun, and P. Michaleris, Prediction of buckling distortion of welded structures. *Journal of Science and Technology in Welding*, 2003. **8**(1): 55–61.
405. Mackerle, J., Finite element analysis and simulation of welding: a bibliography (1976–1996). *Modelling and Simulation in Materials Science and Engineering*, 1996. **4**(5): 501–533.
406. Mackerle, J., Finite element analysis and simulation of welding – an addendum: a bibliography (1996–2001). *Modelling and Simulation in Materials Science and Engineering*, 2002. **10**(3): 295–318.
407. Goldak, J., *et al.*, Predicting distortion and residual stress in complex welded structures by designers. In *7th International Trends in Welding Research*. 2005. Pine Mountain, Georgia, USA.
408. Lu, F., *et al.*, Modeling and finite element analysis on GTAW arc and weld pool. *Computational Materials Science*, 2004. **29**(3): 371–378.
409. Hongyuan, F., *et al.*, New general double ellipsoid heat source model. *Science and Technology of Welding and Joining*, 2005. **10**(3): 361–368.
410. Thiessen, R.G. and I.M. Richardson, A physically based model for microstructure development in a macroscopic heat-affected zone: grain growth and recrystallization. *Metallurgical and Materials Transactions*, 2006. **37B**(4): 655.
411. Thiessen, R., J. Sietsma, and I. Richardson, Microstructural modelling of the HAZ in dual-phase steel. In *8th International Seminar Numerical Analysis of Weldability*. 2006. Seggau, Austria.
412. Elmer, J., *et al.*, Kinetic modeling of phase transformations occurring in the HAZ of C-Mn steel welds based on direct observations. *Acta Materialia*, 2003. **51**: 3333–3349.
413. Zhang, W., J. Elmer, and T. DebRoy, Integrated modelling of thermal cycles, austenite formation, grain growth and decomposition in the heat affected zone of carbon steel. *Science and Technology of Welding and Joining*, 2005. **10**(5): 574–582.
414. Bhadeshia, H.K.D.H., L.E. Svensson, and B. Gretoft, A model for the development of microstructure in low-alloy steel (Fe-Mn-Si-C) weld deposits. *Acta Metallurgica*, 1985. **33**(7): 1271–1283.
415. Elmer, J.W., *et al.*, *In situ* observations of lattice expansion and transformation rates of [alpha] and [beta] phases in Ti-6Al-4V. *Materials Science and Engineering A*, 2005. **391**(1–2): 104–113.
416. Jensen, M., *et al.*, Residual stresses in a welded superalloy disc: characterisation using synchrotron diffraction and numerical process modeling. *Metallurgical and Materials Transactions A*, 2002. **33A**: 2921–2931.
417. Bergheau, J.-M., *et al.*, Viscoplastic behaviour of steels during welding. *Science and Technology of Welding and Joining*, 2004. **9**(4): 323–330.
418. Wilson, E., K.-J. Bathe, and F. Peterson, Finite element analysis of linear and nonlinear heat transfer. *Nuclear Engineering and Design*, 1974. **29**: 110–124.
419. Preston, R.V., *et al.*, Physically-based constitutive modelling of residual stress development in welding of aluminium alloy 2024. *Acta Materialia*, **52**(17): 4973–4983.
420. Barsoum, Z., *Residual Stress Analysis and Fatigue of Welded Structures*. 2006: Department of Aeronautical and Vehicle Engineering, Royal Institute of Technology, Stockholm.

421. Wei, Y.H., *et al.*, Three-dimensional numerical simulation of weld solidification cracking. *Modelling and Simulation in Materials Science and Engineering*, 2005. **13**(3): 437–454.
422. Dye, D., O. Hunziker, and R. Reed, Numerical analysis of the weldability of superalloys. *Acta Materialia, 2001.* **49**: 683–697.
423. Teng, T.-L., C.-P. Fung, and P.-H. Chang, Effect of residual stresses on the fatigue of butt joints using thermal elasto-plastic and multiaxial fatigue theory. *Engineering Failure Analysis*, 2003. **10**: 131–151.
424. Radaj, D., Integrated finite element analysis of welding residual stress and distortion. In *Mathematical Modelling of Weld Phenomena.* 2001. Graz, Austria: Institute of Materials, Minerals and Mining.
425. Teng, T.-L., C.-P. Fung, and P.-H. Chang, Effect of weld geometry and residual stresses on fatigue in butt-welded joints. *International Journal of Pressure Vessels and Piping*, 2002. **79**(7): 467–482.
426. Duranton, P., *et al.*, 3D modelling of multipass welding of a 316L stainless steel pipe. *Journal of Materials Processing Technology*, 2004. **153–154**: 457–463.
427. Audronis, M. and J. Bendikas, The welding deformations of chrome–nickel stainless steels. *Materials Science*, 2003. **9**(2): 22–26.
428. Camilleri, D. and T.G.F. Gray, Computationally efficient welding distortion simulation techniques. *Modelling and Simulation in Materials Science and Engineering*, 2005. **13**(8): 1365.
429. Camilleri, D., P. Mollicone, and T.G.F. Gray, Alternative simulation techniques for distortion of thin plate due to fillet-welded stiffeners. *Modelling and Simulation in Materials Science and Engineering*, 2006. **14**(8): 1307–1327.
430. Deo, M.V. and P. Michaleris, Mitigation of welding induced buckling distortion using transient thermal tensioning. *Science and Technology of Welding and Joining*, 2003. **8**(1): 49–63.
431. van der Aa, E., M. Hermans, and I. Richardson, Conceptual model for stress and strain development during welding with trailing heat sink. *Science and Technology of Welding and Joining*, 2006. **11**(4): 488–495.
432. Toyoda, M. and M. Mochizuki, Control of mechanical properties in structural steel welds by numerical simulation of coupling among temperature, microstructure, and macro-mechanics. *Science and Technology of Advanced Materials*, 2004. **5**(1–2): 255–266.
433. Mcdonald, E.J., K.R. Hallam, and P.E.J. Flewitt, A strategy for accommodating residual stresses in the assessment of repair weldments based upon measurement of near surface stresses. *International Journal of Pressure Vessels and Piping*, 2005. **82**(4): 339–346.
434. Elcoate, C.D., *et al.*, Three-dimensional multi-pass repair weld simulations. *International Journal of Pressure Vessels and Piping*, 2005. **82**(4): 244–257.
435. Vinas, G., *et al.*, Finite element analysis of residual stresses in a letter-box excavation repair weld for Chapelcross Power Station. *International Journal of Pressure Vessels and Piping*, 2005. **82**(4): 270–278.
436. Song, J., J.Y. Shanghvi, and P. Michaleris, Sensitivity analysis and optimization of thermo-elasto-plastic processes with applications to welding side heater design. *Computer Methods in Applied Mechanics and Engineering*, 2004. **193**(42–44): 4541–4566.
437. Voutchkov, I., *et al.*, Weld sequence optimization: the use of surrogate models for solving sequential combinatorial problems. *Computer Methods in Applied Mechanics and Engineering*, 2005. **194**(30–33): 3535–3551.

438. Berglund, D., H. Alberg, and H. Runnemalm, Simulation of welding and stress relief heat treatment of an aero engine component. *Finite Elements in Analysis and Design*, 2003. **39**(9): 865–881.
439. Cho, J.R., *et al.*, Investigation of residual stress and post weld heat treatment of multi-pass welds by finite element method and experiments. *Journal of Materials Processing Technology*, 2004. **155–156**: 1690–1695.
440. Thorborg, J., J. Hald, and J. Hattel, *Stellite failure on a P91 HP Valve – Failure Investigation and Modelling of Residual Stress*. 2004: IIW.
441. Mackerle, J., Finite element analysis and simulation of quenching and other heat treatment processes: a bibliography (1976–2001). *Computational Materials Science*, 2003. **27**(3): 313–332.

Index

FSC
www.fsc.org
MIX
Papier | Fördert gute Waldnutzung
FSC® C083411

Kolibri 360 GmbH
Ferdinand-Jühlke-Straße 7
99095 Erfurt, Deutschland
produktsicherheit@kolibri360.de